Fundamentals of Ceramics

Fundamentals of Ceramics

Anastasia Stokes

Larsen & Keller
www.larsen-keller.com

Fundamentals of Ceramics
Anastasia Stokes
ISBN: 978-1-64172-397-8 (Hardback)

Larsen & Keller

Published by Larsen and Keller Education,
5 Penn Plaza,
19th Floor,
New York, NY 10001, USA

Cataloging-in-Publication Data

Fundamentals of ceramics / Anastasia Stokes.
   p. cm.
Includes bibliographical references and index.
ISBN 978-1-64172-397-8
1. Ceramics. 2. Building materials. I. Stokes, Anastasia.
TP807 .F86 2020
620.14--dc23

For more information regarding Larsen and Keller Education and its products, please visit the publisher's website www.larsen-keller.com

Table of Contents

Preface

It is with great pleasure that I present this book. It has been carefully written after numerous discussions with my peers and other practitioners of the field. I would like to take this opportunity to thank my family and friends who have been extremely supporting at every step in my life.

Ceramic is a material that consists of an inorganic compound of metal or non-metal atoms that are held in ionic and covalent bonds. Some of the physical properties which are generally associated with them are hardness, brittleness, resistance to chemical erosion and ability to withstand very high temperatures. They are broadly divided on the basis of their structure into crystalline ceramics and noncrystalline ceramics. They can also be classified as oxides, non-oxides and composite materials. Some of the modern ceramic materials are aluminum oxide, silicon carbide and tungsten carbide. There are numerous applications of ceramics such as in carbon-ceramic brake disks, ballistic armored vests and dental implants. The topics included in this book on ceramic materials are of utmost significance and bound to provide incredible insights to readers. Different approaches, evaluations and advanced studies on ceramic materials have been included in it. The coherent flow of topics, student-friendly language and extensive use of examples make this book an invaluable source of knowledge.

The chapters below are organized to facilitate a comprehensive understanding of the subject:

Chapter – What is Ceramics?

The solid materials which are made up of an inorganic compound comprising of metal, non-metal or metalloid atoms are known as ceramics. They can be generally classified as traditional ceramics and advanced ceramics. All the diverse aspects of these types of ceramics have been briefly introduced in this chapter.

Chapter – Classification of Ceramics

Ceramics are classified into various categories such as clay products, glass-ceramics, refractories, porcelain, ceramic-impregnated fabrics, terracotta, optical ceramics, abrasives, ceramics cements, and transparent ceramics. This chapter closely examines these categories of ceramics to provide an extensive understanding of the subject.

Chapter – Ceramic Production Techniques

There are numerous techniques which are used to produce ceramics such as sintering, slip casting, freeze-casting, compaction of ceramic powders, ceramic mold casting, ceramic matrix composite and ceramic shell molding. This chapter discusses in detail these production techniques related to ceramics.

Chapter – Ceramography

The branch of science which is involved in creating objects from non-metallic, inorganic materials is known as ceramic engineering. It also deals with testing the fracture toughness, hardness and flexural strength of ceramics. The topics elaborated in this chapter will help in gaining a better perspective about these applications of ceramic engineering.

Chapter – Applications of Ceramics

Ceramics are applied in a wide range of fields such as electronics, dentistry, art, transportation, high pressure fuel systems, thermal insulators, chemical industry, pharmaceutical industry and automotive engineering. The diverse applications of ceramics in these fields have been thoroughly discussed in this chapter.

Anastasia Stokes

1

What is Ceramics?

The solid materials which are made up of an inorganic compound comprising of metal, non-metal or metalloid atoms are known as ceramics. They can be generally classified as traditional ceramics and advanced ceramics. All the diverse aspects of these types of ceramics have been briefly introduced in this chapter.

Ceramics are special materials with many applications in almost all the engineering disciplines. But their importance has often been underestimated due to the fact that many people believe that ceramics are all about pottery and tiles. Today's ceramics industry is one of most rapidly advancing concerns in many parts of the world. Ceramic industry began to expand as a modern industry with the attribution of new techniques and knowledge gained in the 1970s. Since then it has also been one of most competitive industries in the market.

Ceramic Materials

Ceramic materials are special because of their properties. They typically possess high melting points, low electrical and thermal conductivity values, and high compressive strengths. Also they are generally hard and brittle with very good chemical and thermal stability. Ceramic materials can be categorized as traditional ceramics and advanced ceramics. Ceramic materials like clay are categorized as traditional ceramics and normally they are made of clay, silica, and feldspar. As its name suggests, traditional ceramics are not supposed to meet rigid specific properties after their production, so cheap technologies are utilized for most of the production processes.

Properties of Ceramics

Ceramics are weak, although the commonly held belief is that ceramics are weak, this can now be dispelled by the range of ceramic materials available from Dynamic-Ceramic. Some of our Technox range has ceramic properties that are well into the regime of metallic materials.

Some of the highly desirable properties of, Technox ceramic alloys, are:

- High Strength.
- High Fracture Toughness.

- High hardness.

- Excellent wear resistance.

- Good frictional behaviour.

- Anti-static.

- Non-magnetic.

- Low thermal conductivity.

- Corrosion resistance in acids and alkalis.

- Excellent surface finish (0.006 µm Ra).

- Modulus of elasticity similar to steel.

- Thermal expansion coefficient similar to cast iron.

Technox ceramics can be used to produce springs and nails. Probably not but it's true due to the properties of Technox ceramics. The inter-atomic bonding of Advanced Ceramics has long been recognised as responsible for the high values of hardness and compressive ceramic strength. These highly desirable properties of ceramics have as yet been largely disregarded, due to the perceived low toughness and brittle failure demonstrated by traditional ceramic strength. However, recent developments have led to a new breed of ceramic materials displaying mechanical properties that were previously considered.

The fracture toughness of advanced ceramics is often measured using an indentation technique. A polished surface of the material under test is indented using a Vickers hardness tester, the cracks which emanate from the corners or the indents are then measured and provide an indication of the toughness of the material.

Typical microstructures of weak and a tough Technox® zirconia are detailed below:

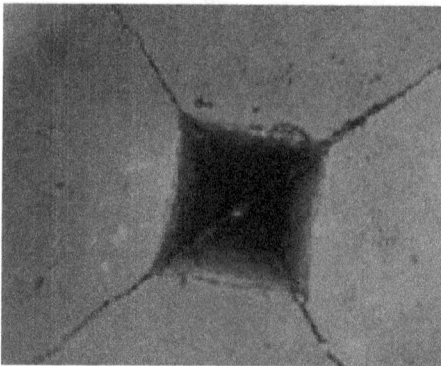

Microstructure of a Low Fracture
Toughness advanced ceramic.

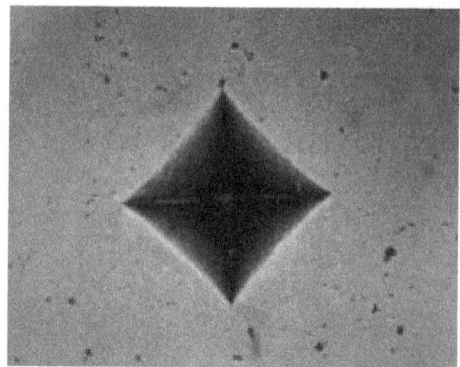

Microstructure of a High Fracture
Toughness advanced ceramic.

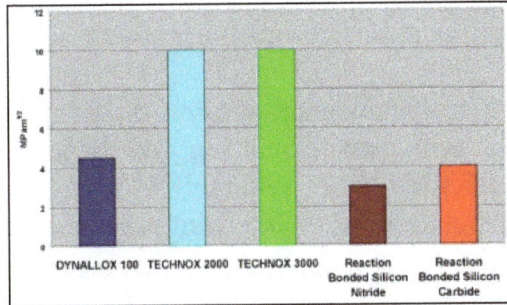

These values are typical of those obtained from test pieces and
are offered for guidance in the design of engineering components.

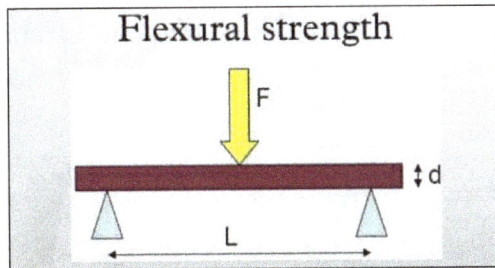

Flexural Strength

Flexural Strength in ceramic materials is usually measured by breaking test bars or
rods in 3 or 4 point bending on a universal testing machine.

The picture to the below shows a bar of Technox® 2000 (300 x 12 x 4mm) being tested
and which is deflecting 4.5mm at its centre. Yet another example of the ultra high
strength of Technox® ceramics.

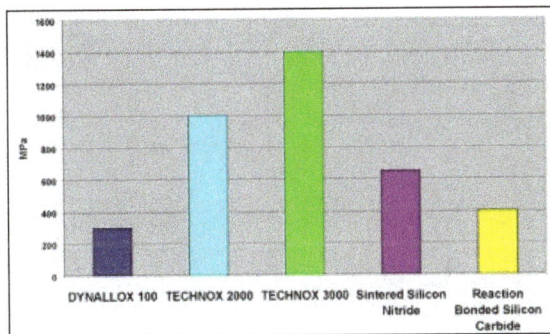

Corrosion Resistance

Ceramics are more resistant to corrosion than most metals and alloys.

Advanced ceramics are highly resistant to chemical corrosion due to their high levels
of chemical stability. Corrosion resistant ceramics possess low chemical solubility and
therefore have particularly high resistance to chemical corrosion. It is their superior

levels of corrosion-resistance that give ceramics the advantage over metallic materials in harsh environments.

In many applications and environments, corrosion-resistant products are a necessity. Harsh chemicals are often used in manufacturing, and even water is corrosive to many. Corrosion resistant ceramics benefit an assembly in terms of effectiveness and productivity and also reduce the need for replacement parts, making them a cost-effective option.

These properties enable our corrosion resistant products to function far better than other materials in corrosive environments.

There are countless applications for corrosion resistant ceramics in a wide variety of industrial and commercial settings where the superior corrosion- and wear-resistance of corrosion resistant products are required.

Thermal Properties

The thermal shock resistance data presented here indicates the maximum thermal downshock that can be tolerated by individual materials.

Thermal downshock (often measured by quenching test pieces directly into cold water) is the most serve problem for advanced ceramics. Thermal upshock causes fewer problems, for example although Technox® 2000 has a downshock limit of ~ 250 °C (depending on sample size and geometry) in upshock it can tolerate a temperature rise of over 800 °C.

Thermal Exansion and Shock Resistance

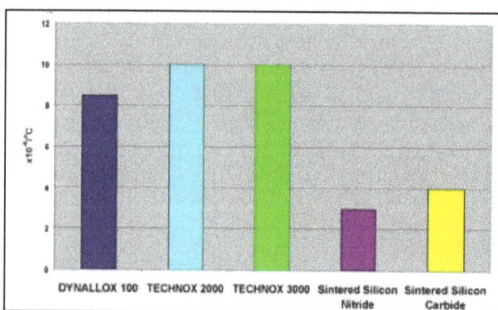

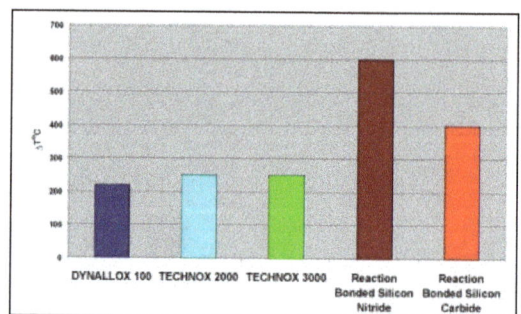

Hardness

The hardness of ceramic materials is a property which is of high significance as it relates to the ability of the material to withstand penetration of the surface through a combination of brittle fracture and plastic flow.

Often, hardness of ceramic material, as with other materials is directly equated to wear resistance. This is a mistaken concept with many metallic components and is definitely an incorrect selection criterion with regards to engineering ceramic materials.

Wear behaviour of ceramic materials is complex and is dependent upon many variables, of which ceramic hardness is an important variable but not the only significant variable.

For example, in many wear environments, such as the erosive wear behaviour of oxide engineering ceramics, it is the ratio of fracture toughness to hardness of ceramics which is found to be of most significance in determining the wear performance.

In many wear environments, a much and "softer" material such as a zirconia can outperform and "harder" materials such as aluminas or silicon carbide.

Ceramic hardness measurements in engineering are generally measured using a Vickers hardness test. In this test a pyramidal diamond indenter is pressed into a polished surface under known loading conditions and the size of the indentation is related to the hardness of the material.

It should also be noted that the hardness value quoted for any material is a function of the type of test conducted and the loading conditions employed. Lighter loads typically provide higher hardness values.

Typically in a Vickers Hardness test, the notation HV10 or HV20 relates to the applied load in kg (in this case 10 or 20 kg respectively).

Other factors that need to be taken into account when interpreting data for hardness of ceramics are the amount of porosity in the surface, the grain size of the micro structure and the effects of grain boundary phases.

Some typical ceramic hardness values are provided below:

Material Class	Vickers Hardness (HV) GPa
Glasses	5 – 10
Zirconias, Aluminium Nitrides	10 – 14
Aluminas, Silicon Nitrides	15 – 20
Silicon Carbides, Boron Carbides	20 – 30
Cubic Boron Nitride CBN	40 – 50
Diamond	60 – 70 >

Wear Behaviour

The wear behaviour of engineering ceramics is relatively complex and is subject to many variables.

Cracking, plastic deformation, tribochemical interaction, abrasion and surface fatigue have all been identified as wear mechanisms operative in ceramic sliding wear situations. The individual ceramic microstructures also affect the wear behaviour in a fundamental manner.

Wear Mechanisms

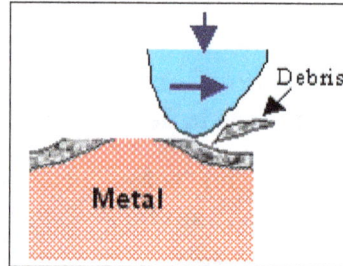

When one considers the intimate contact of two sliding surfaces where hard particles are either present or formed during sliding, abrasive wear can occur as a consequence of both plastic deformation and fracture mechanisms.

However, in polycrystalline ceramics, the amount of plastic deformation that can occur is strictly limited by the available slip systems and twinning modes. Consequently, abrasive wear is aided by fracture mechanisms initiated by the inelastic structure of the material.

Also, on a more microscopic scale than the cracking shown in figure, the intersection of slip bands or twins with barriers such as grain boundaries, particles or other slip bands, can commonly lead to stresses which often give rise to crack nucleation and growth.

Although plastic deformation and fracture have been observed to result in material removal during the abrasive wear of brittle solids, the predominant and rate controlling mechanism differs for both different wear environments and different materials.

Wear Mechanism	Contributory Factors
Microfracture (Trans and Intergranular) Surface Cracking	• Stress Concentration • Second Phases • Flaws • High Young's Modulus • Low Fracture Toughness
Microfracture (Trans and Intergranular) Subsurface Cracking	• Residual Stresses • Inclusions • Flaws • Second Phases • Low Fracture Toughness

Delamination Fracture (Subsurface Cracking Due to Fatigue)	• Plastic deformed layer • Tribochemical reaction layer • Residual stresses • Inclusions, second phases • Flaws
Tribochemical Reaction	• Surface Layers • Stress Corrosion (Environment) • Surfaces Effects (Rehbinder) • Oxidation • Sliding Velocity
Microfracture (Trans and Intergranular) Surface Cracking	• Surface Softening • Plastic Deformation • Structural Changes (crystal structure) • Thermal Shock Cracking • Sliding Velocity
Microfracture (Trans and Intergranular) Surface Cracking	• Transferred Material (adhesion, roughness) • Loose Wear Debris • Compacted Wear Debris
Microfracture (Trans and Intergranular) Surface Cracking	• Microcutting, Microploughing • Microfatigue • Microcracking • Spalling

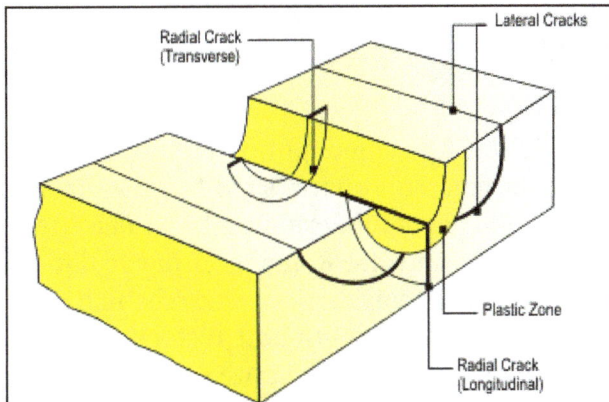

Figure: Crack Modes in Polycrystalline Ceramics

Plastic deformation is favoured when the load on the abrasive particles is small. This occurs as a result of small abrasive particles or low applied loads, when the abrasive is blunt or blunts during contact, and when the ratio of fracture toughness to hardness is high.

Conversely, indentation fracture is favoured when the load on the abrasive particles is high. As occurs with large particles or high applied loads, when the abrasive is sharp or remains sharp due to fracture on contact, and when the ratio of fracture toughness to hardness is low.

This ratio of fracture toughness to hardness has also been shown to be of significance in the erosive wear behaviour of zirconia ceramics. The high value for this ratio with Technox Zirconia ceramics leads to their excellent erosive wear resistance in applications such as pumps and choke valves.

TRADITIONAL CERAMICS

Traditional ceramics are ceramic materials that are derived from common, naturally occurring raw materials such as clay minerals and quartz sand. Through industrial processes that have been practiced in some form for centuries, these materials are made into such familiar products as china tableware, clay brick and tile, industrial abrasives and refractory linings, and portland cement.

Traditional ceramic objects are almost as old as the human race. Naturally occurring abrasives were undoubtedly used to sharpen primitive wood and stone tools, and fragments of useful clay vessels have been found dating from the Neolithic Period, some 10,000 years ago. Not long after the first crude clay vessels were made, people learned how to make them stronger, harder, and less permeable to fluids by burning. These advances were followed by structural clay products, including brick and tile. Clay-based bricks, strengthened and toughened with fibres such as straw, were among the earliest composite materials. Artistic uses of pottery also achieved a high degree of sophistication, especially in China, the Middle East, and the Americas.

With the advent of the Metal Age some 5,000 years ago, early smiths capitalized on the refractory nature of common quartz sand to make molds for the casting of metals—a practice still employed in modern foundries. The Greeks and Romans developed lime-mortar cements, and the Romans in particular used the material to construct remarkable civil engineering works, some of which remain standing to this day. The Industrial Revolution of the 18th and 19th centuries saw rapid improvements in the processing of ceramics, and the 20th century saw a growth in the scientific understanding of these materials. Even in the age of modern advanced ceramics, traditional ceramic products, made in large quantities by efficient, inexpensive manufacturing methods, still make up the bulk of ceramics sales worldwide. The scale of plant operations can rival those found in the metallurgical and petrochemical industries.

Raw Materials

Because of the large volumes of product involved, traditional ceramics tend to be manufactured from naturally occurring raw materials. In most cases these materials are silicates—that is, compounds based on silica (SiO_2), an oxide form of the element silicon. In fact, so common is the use of silicate minerals that traditional ceramics are often referred to as silicate ceramics, and their manufacture is often called the silicate industry. Many of the silicate materials are actually unmodified or chemically modified aluminosilicates (alumina [Al_2O_3] plus silica), although silica is also used in its pure form. Altogether, the raw materials employed in traditional ceramics fall into three commonly recognized groups: Clay, silica, and feldspar. These groups are described below:

Clay

Clay minerals such as kaolinite ($Al_2[Si_2O_5][OH]_4$) are secondary geologic deposits, having been formed by the weathering of igneous rocks under the influence of water, dissolved carbon dioxide, and organic acids. The largest deposits are believed to have formed when feldspar ($KAlSi_3O_8$) was eroded from rocks such as granite and was deposited in lake beds, where it was subsequently transformed into clay.

The importance of clay minerals to traditional ceramic development and processing cannot be overemphasized. In addition to being the primary source of aluminosilicates, these minerals have layered crystal structures that result in plate-shaped particles of extremely small micrometre size. When these particles are suspended in or mixed with water, the mixture exhibits unusual rheology, or flow under pressure. This behaviour allows for such diverse processing methods as slip casting and plastic forming, which are described below. Clay minerals are therefore considered to be formers, allowing the mixed ingredients to be formed into the desired shape.

Silica and Feldspar

Other constituents of traditional ceramics are silica and feldspar. Silica is a major ingredient in refractories and whitewares. It is usually added as quartz sand, sandstone, or flint pebbles. The role of silica is that of a filler, used to impart "green" (that is, unfired) strength to the shaped object and to maintain that shape during firing. It also improves final properties. Feldspars are aluminosilicates that contain sodium (Na), potassium (K), or calcium (Ca). They range in composition from $NaAlSi_3O_8$ and $KAlSi_3O_8$ to $CaAl_2Si_2O_8$. Feldspars act as fluxing agents to reduce the melting temperatures of the aluminosilicate phases.

Processing

Beneficiation

Compared with other manufacturing industries, far less mineral beneficiation (e.g., washing, concentrating, sizing of particulates) is employed for silicate ceramics. Clays going into common structural brick and tile are often processed directly as dug out of the ground, although there may be some blending, aging, and tempering for uniform distribution in water. Such impure clays are workable in untreated form because they already contain fillers and fluxes in association with the clay minerals. In the case of whitewares, for which the raw materials must be in a purer state, the clays are washed, and impurities are either settled out or floated off. Silicas are purified by washing and separating unwanted minerals by gravity and by magnetic and electrostatic means. Feldspars are beneficiated by flotation separation, a process in which a frothing agent is added to separate the desired material from impurities.

Blending

The calculation of amounts, weighing, and initial blending of raw materials prior to forming operations is known as batching. Batching has always constituted much of the art of the ceramic technologist. Formulas are traditionally jealously guarded secrets, involving the selection of raw materials that confer the desired working characteristics and responses to firing and that yield the sought-after character and properties. Clays must be selected on the basis of workability, fusibility, fired colour, and other requirements. Silicas, likewise, must meet criteria of chemical purity and particle size distribution.

Forming

The fine, platy morphology of clay particles is used to advantage in the forming of clay-based ceramic products. Depending upon the amount of water added, clay-water bodies can be stiff or plastic. Plasticity arises by virtue of the plate-shaped clay particles slipping over one another during flow. (Nonclay ceramics can be similarly formed if plasticizers—usually polymers—are added to their mixes. In many cases organic binders are used to help hold the body together until it is fired). With even higher water content and the addition of dispersing agents to keep the clay particles in suspension, readily flowable suspensions can be produced. These suspensions are called slips or slurries and are employed in the slip casting of clay bodies.

Plastic Forming

Plastic forming is the primary means of shaping clay-based ceramics. After the raw materials are mixed and blended into a stiff mud or plastic mix, a variety of forming techniques are employed to produce useful shapes, depending upon the ceramic involved and the type of product desired. Foremost among these techniques are pressing and extrusion.

Pressing involves the application of pressure to eliminate porosity and achieve a specific shape, depending upon the die employed. Refractory bricks, for example, are often made by die presses that are either single-action (pressing from the top only) or dual-action (simultaneously pressing from top and bottom). Structural clay products such as brick and tile can be made in the same fashion. In pressing operations the feed material tends to have a lower water content and is referred to as a stiff mud.

The problem with die casting is that it is a piecemeal rather than a continuous process, thereby limiting throughput. Many silicate ceramics are therefore manufactured by extrusion, a process that allows a more efficient continuous production. In a commercial screw-type extruder, a screw auger continuously forces the plastic feed material through an orifice or die, resulting in simple shapes such as cylindrical rods and pipes, rectangular solid and hollow bars, and long plates. These shapes can be cut upon extrusion into shorter pieces for bricks and tiles.

Slip Casting

A different approach to the forming of clay-based ceramics is taken in slip casting of whiteware, as shown in Figure. As mentioned above, with sufficient water content and the addition of suitable dispersing agents, clay-water mixtures can be made into suspensions called slurries or slips. These highly stable suspensions of clay particles in water arise from the careful manipulation of surface charges on the platelike clay particles. Without a dispersing agent, oppositely charged edges and surfaces of the particles would attract, leading to flocculation, a process in which groups of particles coagulate into flocs with a characteristic house-of-cards structure. Dispersing agents neutralize some of the surface charges, so that the particles can be made to repel one another and remain in suspension indefinitely. When the suspension is poured into a porous plaster mold, capillary forces suck the water into the mold from the slip and cause a steady deposition of clay particles, in dense face-to-face packing, on the inside surface of the mold. After a sufficient thickness of deposit has been obtained, the remaining slip can be poured off or drained and the mold opened to reveal a freestanding clay piece that can be dried and fired. Surprisingly complex shapes can be achieved through slip casting.

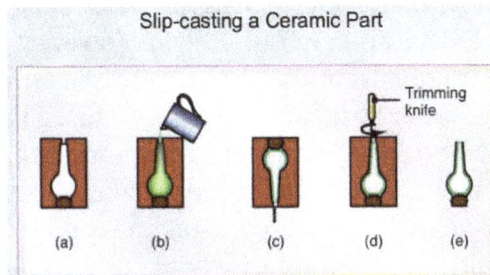

Stages in the slip casting of a thin-walled whiteware container.

Clay powder is mixed in water together with a dispersing agent, which keeps the clay particles suspended evenly throughout the clay-water slurry, or slip. The slip is poured into a plaster mold, where water is drawn out by capillary action and a cast is formed by the deposition of clay particles on the inner surfaces of the mold. The remaining slip is drained, and the cast is allowed to dry partially before the drain hole is plugged and the mold separated. The unfinished ware is given a final drying in an oven before it is fired into a finished product.

Firing

Kiln Operation

After careful drying to remove evaporable water, clay-based ceramics undergo gradual heating to remove structural water, to decompose and burn off any organic binders used in forming, and to achieve consolidation of the ware. Batches of specialty products, produced in smaller volumes, are cycled up and down in so-called batch furnaces. Most mass-produced traditional ceramics, on the other hand, are fired in tunnel kilns. These consist of continuous conveyor belt or railcar operations, with the ware traversing the kiln and gradually being heated from room temperature, through a hot zone, and back down to room temperature. Pyrometric cones, which deform and sag at specific temperatures, often ride with the ware to monitor the highest temperature seen in the traverse through the kiln.

Vitrification

The ultimate purpose of firing is to achieve some measure of bonding of the particles (for strength) and consolidation or reduction in porosity (*e.g.,* for impermeability to fluids). In silicate-based ceramics, bonding and consolidation are accomplished by partial vitrification. Vitrification is the formation of glass, accomplished in this case through the melting of crystalline silicate compounds into the amorphous, noncrystalline atomic structure associated with glass. As the formed ware is heated in the kiln, the clay component turns into progressively larger amounts of glass. The partial vitrification process can be analyzed through a phase diagram such as that shown in Figure. In this diagram three crystalline phases are shown: The end members cristobalite (one crystallographic form of silica [SiO_2]) and alumina(Al_2O_3) and an intermediate compound, mullite ($3Al_2O_3 \cdot 2SiO_2$). The melting points of alumina and cristobalite, as shown on the left and right edges of the diagram, are quite high. However, intermediate compositions begin to melt at lower temperatures. As shown by the two horizontal lines on the diagram, melting begins to occur at 1,828 °C (3,322 °F) for high alumina compositions and as low as 1,587 °C (2,889 °F) for high silica compositions. (These temperatures can be lowered still further by the addition of fluxing agents, such as alkali or alkaline-earth oxide feldspars). Between the two horizontal lines and the region of the diagram marked liquid, all compositions are only partly liquid (*e.g.,* mullite and liquid, alumina and liquid). This partial vitrification allows for the retention of solid particles, which helps to maintain the rigidity of the ceramic piece during firing in order to minimize sagging or warpage.

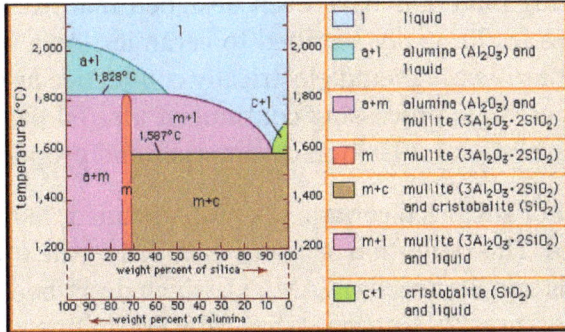

Phase diagram of the alumina-silica system.

Depending on the temperature and on the content of silica and alumina, aluminosilicate clays, upon heating, form various combinations of alumina, cristobalite, mullite, and liquid. The formation of liquid phases is important in the partial vitrification of clay-based ceramics.

The role of the glassy liquid phase in the consolidation of fired clay objects is to facilitate liquid-phase or reactive-liquid sintering. In these processes the liquid first brings about a denser rearrangement of particles by viscous flow. Second, through solution-precipitation of the solid phases, small particles and surfaces of larger particles dissolve and reprecipitate at the growing "necks" that connect large particles. Rearrangement and solution-precipitation lead to bond formation and to progressive densification with reduction of porosity. A range of glass contents and residual porosities can be obtained, depending on the ingredients and the time the object is held at maximum temperature.

Finishing

If fired ceramic ware is porous and fluid impermeability is desired, or if a purely decorative finish is desired, the product can be glazed. In glazing, a glass-forming formulation is pulverized and suspended in an appropriate solvent. The fired ceramic body is dipped in or painted with the glazing slurry, and it is refired at a temperature that is lower than its initial firing temperature but high enough to vitrify the glaze formulation. Glazes can be coloured by the addition of specific transition-metal or rare-earth elements to the glaze glass or by the suspension of finely divided ceramic particles in the glaze.

ADVANCED CERAMICS

Advanced ceramics are substances and processes used in the development and manufacture of ceramic materials that exhibit special properties.

Advanced ceramics represent an "advancement" over this traditional definition. Through the application of a modern materials science approach, new materials or new combinations of existing materials have been designed that exhibit surprising variations on the properties traditionally ascribed to ceramics. As a result, there are now ceramic products that are as tough and electrically conductive as some metals. Developments in advanced ceramic processing continue at a rapid pace, constituting what can be considered a revolution in the kind of materials and properties obtained.

With the development of advanced ceramics, a more detailed, "advanced" definition of the material is required. This definition has been supplied by the 1993 Versailles Project on Advanced Materials and Standards (VAMAS), which described an advanced ceramic as "an inorganic, nonmetallic (ceramic), basically crystalline material of rigorously controlled composition and manufactured with detailed regulation from highly refined and/or characterized raw materials giving precisely specified attributes." A number of distinguishing features of advanced ceramics are pointed out in this definition. First, they tend to lack a glassy component; i.e., they are "basically crystalline." Second, microstructures are usually highly engineered, meaning that grain sizes, grain shapes, porosity, and phase distributions (for instance, the arrangements of second phases such as whiskers and fibres) are carefully planned and controlled. Such planning and control require "detailed regulation" of composition and processing, with "clean-room" processing being the norm and pure synthetic compounds rather than naturally occurring raw materials being used as precursors in manufacturing. Finally, advanced ceramics tend to exhibit unique or superior functional attributes that can be "precisely specified" by careful processing and quality control. Examples include unique electrical properties such as superconductivity or superior mechanical properties such as enhanced toughness or high-temperature strength. Because of the attention to microstructural design and processing control, advanced ceramics often are high value-added products.

Advanced ceramics are referred to in various parts of the world as technical ceramics, high-tech ceramics, and high-performance ceramics. The terms engineering ceramics and fine ceramics are used in the United Kingdom and Japan, respectively.

Chemical Routes to Precursors

Like their traditional counterparts, advanced ceramics are often made by mixing and calcining (firing together) precursor powders. Unlike traditional ceramics, however, naturally occurring raw materials are seldom employed. Instead, highly pure synthetic precursors are typically used. In addition, liquid-phase sintering, a method of densifying powders that is common in traditional ceramic processing, is seldom employed. Instead, advanced ceramics are densified by transient-liquid sintering (also referred to as reactive-liquid sintering) or solid-state sintering. The most important factor in these sintering methods is small particle size. Small particles have a larger ratio of surface area to mass and therefore yield a higher driving force for sintering. Small particle sizes also reduce the distances over which diffusion of material must take place. Ceramists therefore take care to produce active ceramic powders with small grain size, usually in the submicrometre range—i.e., smaller than one micrometre, or one-millionth of a metre (0.000039 inch).

A major issue in the preparation of powdered precursors, especially for electroceramic applications, is chemical homogeneity—that is, the establishment of uniform chemical composition throughout the mixture. Standard solid-state techniques for processing separate precursor powders can approach homogeneity in the final product only after many grinding and firing steps. A number of chemical approaches therefore have been developed in order to improve mixing, even down to the atomic level. Often these techniques involve the decomposition of salts—for instance, carbonates, nitrates, and sulfates—into the desired chemical form. Most ceramics, are oxides of metallic elements, although many ceramics (especially advanced ceramics) consist of carbide, nitride, and boride compounds as well. The various chemical techniques for achieving homogenous, small-grained powders are described.

Coprecipitation and Freeze-drying

Often the salt compounds of two desired precursors can be dissolved in aqueous solutions and subsequently precipitated from solution by pH adjustment. This process is referred to as coprecipitation. With care, the resulting powders are intimate and reactive mixtures of the desired salts. In freeze-drying, another route to homogenous and reactive precursor powders, a mixture of water-soluble salts (usually sulfates) is dissolved in water. Small droplets are then rapidly frozen by spraying the solution into a chilled organic liquid such as hexane. With rapid freezing of the spray droplets into small ice crystals, segregation of the chemical constituents is minimized. The frozen material is removed from the hexane by sieving, and water is then removed from the ice by sublimation under vacuum.

After coprecipitation or freeze-drying the resulting powders undergo intermediate high-temperature calcination to decompose the salts and produce fine crystallites of the desired oxides.

Spray Roasting

Spray roasting involves spray atomization of solutions of water-soluble salts into a heated chamber. The temperature and transit time are adjusted so as to accomplish rapid evaporation and oxidation. The result is a high-purity powder with fine particle size. A modification of spray roasting, known as rapid thermal decomposition of solutions (RTDS), can yield nano-size oxide powders—that is, particles measured in nanometres (one-billionth of a metre).

The Sol-gel Route

An increasingly popular method for producing ceramic powders is sol-gel processing. Stable dispersions, or sols, of small particles (less than 0.1 micrometre) are formed from precursor chemicals such as metal alkoxides or other metalorganics. By partial evaporation of the liquid or addition of a suitable initiator, a polymer-like, three-dimensional bonding takes place within the sol to form a gelatinous network, or gel. The gel can then be dehydrated and calcined to obtain a fine, intimately mixed ceramic powder.

The Pechini Process

A process related to the sol-gel route is the Pechini, or liquid mix, process (named after its American inventor, Maggio Pechini). An aqueous solution of suitable oxides or salts is mixed with an alpha-hydroxycarboxylic acid such as citric acid. Chelation, or the formation of complex ring-shaped compounds around the metal cations, takes place in the solution. A polyhydroxy alcohol is then added, and the liquid is heated to 150–250 °C (300–480 °F) to allow the chelates to polymerize, or form large, cross-linked networks. As excess water is removed by heating, a solid polymeric resin results. Eventually, at still higher temperatures of 500–900 °C (930–1,650 °F), the resin is decomposed or charred, and ultimately a mixed oxide is obtained. Particle size is extremely small, typically 20 to 50 nanometres (although there is agglomeration of these particles into larger clusters), with intimate mixing taking place on the atomic scale.

Combustion Synthesis

A modification of the Pechini process is combustion synthesis. One version of this process involves a reaction between nitrate solutions and the amino acid glycine. The glycine, in addition to complexing with the metal cations and increasing their solubility, serves as a fuel during charring. After much of the water has been evaporated, a viscous liquid forms that autoignites around 150–200 °C (300–400 °F). Combustion temperatures rapidly exceed 1,000 °C (1,800 °F) and convert the material to fine, intimately mixed, and relatively nonagglomerated crystallites of the complex oxide desired. This technique is referred to as the glycine-nitrate process.

High-temperature Synthesis

In a reaction known as self-propagating high-temperature synthesis (SHS), highly reactive metal particles ignite in contact with boron, carbon, nitrogen, and silica to form boride, carbide, nitride, and silicide ceramics. Since the reactions are extremely exothermic (heat-producing), the reaction fronts propagate rapidly through the precursor powders. Usually, the ultimate particle size can be controlled by the particle size of the precursors.

Exotic Energy Deposition

So-called exotic energy deposition systems also are employed in the processing of ceramic powders, often resulting in extremely small clusters of atoms or ions or nano-size particles. Among other techniques, vacuum evaporation/condensation can be employed to make nanoparticles. In this system metal sources are heated through electrical resistivity under conditions of ultra-high vacuum. Metal atoms evaporate and then form clusters that deposit on a thermal convection collector that is chilled by liquid nitrogen. The nanoparticles can be oxidized before or after being scraped from the collector.

Consolidation

Traditional Methods

Many of the same consolidation processes used for traditional ceramics—e.g., pressing, extrusion, slip casting—are also employed for advanced ceramics. A high degree of sophistication has been obtained in these processes. An outstanding example is the extruded honeycomb-shaped structure used as the catalyst support in automotive catalytic convertors. These small structures have hundreds of open cells per square centimetre, with wall thicknesses of less than 0.1 millimetre. The extrusion of such fine shapes is made possible by the addition of a hydraulic (water-setting) polymer resin to the mix. The resin rapidly cures upon extrusion of the mix into hot water and imparts "green" (prefired) strength to the structure.

Tape Casting

Tape casting is another process that was originally used with traditional ceramics but has achieved a high level of sophistication for advanced ceramics. In particular, tape-casting methods are used to make substrates for integrated circuits and the multilayer structures used in both integrated-circuit packages and multilayer capacitors. A common tape-casting method is called doctor blading. In this process a ceramic powder slurry, containing an organic solvent such as ethanol and various other additives (e.g., polymer binder), is continuously cast onto a moving carrier surface made of a smooth, "no-stick" material such as Teflon. A smooth knife edge spreads the slurry to a specified thickness, the solvent is evaporated, and the tape is rolled onto a take-up reel for additional processing.

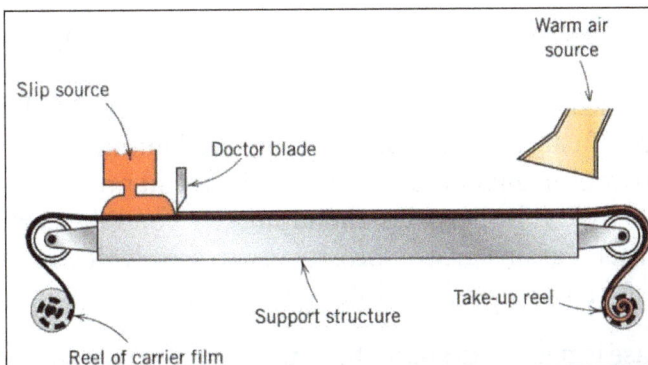

Steps in doctor blading, a tape-casting process
employed in the production of ceramic films.

Ceramic powder and solvent are mixed to form a slurry, which is treated with various additives and binders, homogenized, and then pumped directly to a tape-casting machine. There the slurry is continuously cast onto the surface of a moving carrier film. The edge of a smooth knife, generally called a doctor blade, spreads the slurry onto the carrier film at a specified thickness, thereby generating a flexible tape. Heat lamps gently evaporate the solvent, and the dry tape is peeled away from the carrier film and rolled onto a take-up reel for additional processing.

Two other tape-casting methods are the waterfall technique and the paper-casting process. In the waterfall technique a conveyor belt carries a flat surface through a continuous, recirculated waterfall of slurry. This method—which is commonly employed to coat candy with chocolate—has also been used to form thin-film dielectrics for capacitors as well as thick-film porous electrodes for fuel cells. The paper-casting process involves dipping a continuous paper tape into a ceramic powder slurry. The coated paper is dried and rolled onto take-up reels. In subsequent firing operations the paper is burned away, leaving the ceramic structure.

Injection Molding

Injection molding, commonly employed in polymer processing, also is employed for advanced ceramics. In injection molding a ceramic mix is forced through a heated tubular barrel by action of a screw or plunger. The mix consists of ceramic powder plus a thermoplastic polymer that softens with heat. The heated barrel of the injection molding machine ensures that the mix will flow under pressure, and the screw or plunger forces the fluid mix into a cold die cavity, where the mass hardens to the shape specified by the cavity. Complex shapes can be achieved in injection molding. Outstanding examples are rotors for turbochargers and rotor blades and stator vanes for gas turbines, made from silicon carbide and silicon nitride.

Densification

Solid-state Sintering

Like traditional ceramics, advanced ceramics are densified from powders by applying heat—a process known as sintering. Unlike traditional ceramics, however, advanced powders are not bonded by the particle-dissolving action of glassy liquids that appear at high temperatures. Instead, solid-state sintering predominates. In this process, matter from adjacent particles, under the influence of heat and pressure, diffuses to "neck" regions that grow between the particles and ultimately bond the particles together. As the boundaries between grains grow, porosity progressively decreases until, in a final stage, pores close off and are no longer interconnected.

Since no glassy phase is needed in solid-state sintering to bond particles, there is no residual glass at the grain boundaries of the resulting dense ceramic that would degrade its properties. As a result, advanced ceramics have improved properties—e.g., strength or conductivity—over their liquid-sintered traditional counterparts, especially at elevated temperatures.

Nevertheless, while pores located at the grain boundaries can be eliminated by grain boundary diffusion, pores left inside the growing grains are extremely difficult to eliminate, no matter how long the object is sintered. For this reason sintering aids are often used to enhance the sintering of advanced ceramics. In reactive-liquid, or transient-liquid, sintering, a chemical additive produces a temporary liquid that facilitates the initial stages of sintering. The liquid is subsequently evaporated, resorbed by the solid particles, or crystallized into a solid.

Solid-state sintering is also aided by chemical additives. A classic example is the sintering of alumina lamp envelopes for sodium-vapour street lights. The lamp envelope must be able to contain the hot sodium discharge, and at the same time it must be transparent, or at least translucent, to visible light. The necessary refractory properties can be found in alumina, but the material does not sinter to translucency, and residual pores that remain within the grains act to scatter light. With magnesia as a sintering aid, however, alumina sinters to translucency. Apparently, magnesia slows the migration of grain boundaries during sintering. Pores remain on these boundaries and are eliminated by grain boundary diffusion. Extremely low porosities can be achieved.

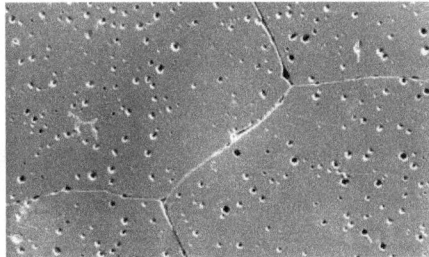

Opaque alumina. In alumina solidified without chemical sintering aids, pores are trapped within the grains, scattering light and contributing to the material's opacity.

Translucent alumina.

With the use of magnesia as an aid during sintering (densifying under heat), light-scattering pores remain on the boundaries between grains and diffuse from the material, helping to make the alumina translucent.

Pressure-assisted Sintering

The sintering processes described above can be assisted by the application of pressure. Pressure increases the driving force for densification, and it also decreases the temperature needed for sintering to as low as half the melting point of the ceramic. Furthermore, shape forming and densification can often be accomplished in a single step. Two popular pressure-assisted sintering methods are followed—hot pressing and hot isostatic pressing (HIP).

In hot pressing a heated single-action or dual-action die press is employed. The material composing or lining the rams and die walls is extremely important, since it must not react with the ceramic being hot-pressed. Unfortunately, complex shapes cannot be processed by hot pressing. Hot isostatic pressing involves immersing the green ceramic in a high-pressure fluid (usually an inert gas such as argon or helium) at elevated temperature. For the applied pressure to squeeze out the residual porosity, the ceramic piece must first be presintered to the closed porosity stage (no open, interconnected pores), or else it must be encapsulated with a viscous coating such as glass. During the "HIPing" process, the high-pressure fluid then presses on the exterior, and residual gases from within the piece bubble out and are eliminated. Preformed complex shapes such as turbine blades, rotors, and stators can be densified by HIP.

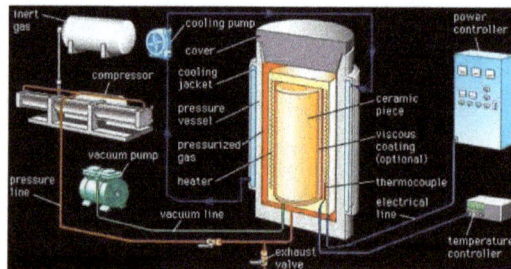

Hot isostatic pressing (HIP), a pressure-assisted
method for sintering advanced ceramic pieces.

A ceramic piece is inserted into the heater compartment of a pressure vessel, which is evacuated of air by means of a vacuum pump. A thermocouple placed between the piece and the heating coils monitors the process temperature, which is regulated by an outside temperature controller. Overall electrical controls are monitored by a computerized power controller. An inert gas is fed under pressure into the vessel; at the end of the HIP cycle the gas is vented through an exhaust valve and the temperature is reduced by cold water pumped through a cooling jacket.

Rapid Heating

Exotic energy deposition methods also are used in the sintering of advanced ceramics. One reason is that conventional radiant heating is slow, so that ceramic powders lose much of their activity, or sinterability, during heat-up. It is therefore advantageous to heat ceramics to the sintering temperature as rapidly as possible. Two means of rapid heating are plasma sintering and microwave sintering. Plasma sintering takes place in an ionized gas. Energetic ionized particles recombine and deposit large amounts of energy on the surfaces of the ceramic being sintered. Extremely high sintering rates have been achieved with this method. In microwave sintering, electromagnetic radiation at microwave frequencies can penetrate and deposit heat in the interior of a sintering ceramic, thus reversing the usual outside-in temperature gradient seen in conventional radiant heating. A combination of radiant and microwave heating can be used to obtain uniform heating throughout the piece. Unfortunately, neither plasma nor microwave sintering is amenable to complex shapes.

Chemical Bonding

Reaction Sintering

Reaction sintering, or reaction bonding, is an important means of producing dense covalent ceramics. Reaction-bonded silicon nitride (RBSN) is made from finely divided silicon powders that are formed to shape and subsequently reacted in a mixed nitrogen/hydrogen or nitrogen/helium atmosphere at 1,200 to 1,250 °C (2,200 to 2,300 °F). The nitrogen permeates the porous body and reacts with the silicon to form silicon nitride within the pores. The piece is then heated to 1,400 °C (2,550 °F), just below the melting point of silicon. Precise control is exercised over the nitrogen flow rate and the heating rate. The entire reaction-sintering process can last up to two weeks. Although up to 60 percent weight gain occurs during nitriding, dimensional change is less than 0.1 percent. This is a "net shape" process, which allows for excellent dimensional control and reduces the amount of costly machining and finishing needed after firing. Since no sintering aids are employed, the high-temperature strength and creep resistance of RBSN are quite good.

Reaction-bonded silicon carbide (RBSC) is produced from a finely divided, intimate mixture of silicon carbide and carbon. Pieces formed from this mixture are exposed to liquid or vapour silicon at high temperature. The silicon reacts with the carbon to form additional silicon carbide, which bonds the original particles together. Silicon also fills any residual open pores. Like RBSN, RBSC undergoes little dimensional change during sintering. Products exhibit virtually constant strength as temperatures rise to the melting point of silicon.

Infiltration

The siliconization of RBSC is a good example of infiltration, which may be described as any technique of filling in pores by reaction with or deposition from a liquid or vapour. In the case of liquid reaction, the technique is called melt infiltration; in the case of vapour phases, it is called chemical vapour infiltration, or CVI. With infiltration it is possible to begin with woven carbon fibres or felts, building up composite materials with enhanced properties.

The Lanxide Process

Another chemical bonding method is the Lanxide process, introduced by the Lanxide Company in the United States. In this process a molten metal is reacted with a gas to form a metal-ceramic composite at the metal-gas interface. As the composite grows at the metal-composite interface, edges remain in contact with the melt and act as a wick for additional reactant metal. The Lanxide process has been employed to produce complex shapes made from such ceramic-metal composites, or cermets, as boron carbide-boron, titanium nitride-titanium, and zirconium boride-zirconium.

Film Deposition

Advanced ceramics intended for electromagnetic and mechanical applications are often produced as thin or thick films. Thick films are commonly produced by paper-casting methods, described above, or by spin-coating. In spin-coating a suspension of ceramic particles is deposited on a rapidly rotating substrate, with centrifugal force distributing the particles evenly over the surface. On the other hand, truly thin films (that is, films less than one micrometre thick) can be produced by such advanced techniques as physical vapour deposition (PVD) and chemical vapour deposition (CVD). PVD methods include laser ablation, in which a high-energy laser blasts material from a target and through a vapour to a substrate, where the material is deposited. Another PVD approach involves sputtering, in which energetic electrons bombard the surface of a target, removing material as a vapour that is deposited on an adjacent substrate. CVD involves passing a carrier gas over a volatile organometallic precursor; the gas and organometallic react, producing a ceramic compound that is deposited downstream on an appropriate substrate.

Even more precise control over the deposition of thin films can be achieved by molecular beam epitaxy, or MBE. In this technique molecular beams are directed at and react with other molecular beams at the substrate surface to produce atomic layer-by-layer deposition of the ceramic. Epitaxy (in which the crystallinity of the growing thin film matches that of the substrate) can often be achieved. Such films have potential for advanced electronic and photonic applications, including superconductivity.

SUPER HARD CERAMICS

In the world of technical ceramics, there are two materials which are only surpassed by diamond in terms of hardness – and both are used by Precision Ceramics as a base material for a wide range of technical components in an equally wide field of applications.

In terms of toughness, there's not much to choose between them but each has its own specific advantages in terms of properties and application.

Boron Carbide, for instance, is currently the hardest material produced in tonnage quantities and is the third hardest material known to man after diamond and cubic boron nitride.

Not far behind it in the hardness stakes comes silicon carbide, more commonly known as carborundum, steeped in history since it was first mass-produced in 1893 and probably the most common of all industrial abrasives.

B_4O

The extreme hardness of boron carbide provides excellent wear and abrasion resistance and consequently it is a perfect base material for the manufacture of nozzles for slurry pumping, grit blasting and in water jet cutters.

In combination with other materials, boron carbide also finds extensive use in ballistic armour (including body and personnel armour) where its combination of high hardness, high elastic modulus, and low density gives the material an exceptionally high specific stopping power to defeat high velocity projectiles.

Other applications include ceramic tooling dies, precision tool parts, evaporating boats for materials testing and mortars and pestles.

SiC

The technical properties of silicon carbide are remarkably similar to those of diamond. It is one of the lightest, hardest and strongest technical ceramic materials and has exceptional thermal conductivity, chemical resistance and low thermal expansion.

Silicon carbide is an excellent material to use when physical wear is an important consideration because it provides good erosion and abrasive resistance making it particularly suitable for such applications as spray nozzles, shot blast nozzles and cyclone components.

Not so long ago, silicon carbide was the chosen material to line the brakes of the most advanced, jaw-dropping car the world has ever seen, the £850K plus McLaren P1. And nowadays you don't have to drive too far through the English countryside to see a wind turbine in which the turbine bearings are almost certainly likely to be made from silicon carbide.

TRADITIONAL VERSUS ADVANCED CERAMICS

Different Raw Materials

Traditional ceramics use natural minerals without processing, such as clay, quartz, feldspar and so on. However, the raw materials of precise ceramics are synthetic high-quality powders, which is a breakthrough in traditional ceramic clay. The "well selected raw materials" endow precise ceramics with more good qualities and functions.

Raw Material

Different Structures

The structure of traditional ceramic is decided by the composition of the clay. The ceramics from different origins have various textures. Because of the using of different raw materials, traditional ceramics tend to have more complicated chemical structures and compositions. Besides, traditional ceramic has more impurities both in type and quantity. The microstructure of traditional ceramics is not even with multiple pores. Therefore, it's harder to control the quality of traditional ceramic products. The chemical structures of precise ceramics are simple and clear with high purity. Moreover, precise ceramics are made from manually calculated ingredients, which means the raw materials are under control. Therefore, the microstructure of advanced ceramics is generally uniform and fine.

Different Manufacturing Processes

The minerals for traditional ceramics can be directly used for wet moldings, such as plastic molding of mud or grouting molding of the slurry. The products need no more processing after sintering with the temperature between 1652 °F to 2552 °F. However, dry molding and wet molding can only be suitable for precise ceramics when organic additions are added to the raw materials of high-purity powders. Precise ceramics still need more processing after firing under a higher sintering temperature, from 2192 °F to 3992 °F according to different materials. From the perspective of preparation procedures, precise ceramics overcome the limits of traditional ceramics. Moreover, there are many advanced technologies used in precise ceramics, such as vacuum sintering, protective atmosphere sintering, hot pressing, hot and high-temperature isostatic pressing, and so on.

Different Functions

With the above differences, traditional ceramics and precise ceramics own different functions. Precise ceramics have better performances in quality as well as new applications that traditional ones don't have. Traditional ceramic materials are mainly produced for daily use or work as building materials. While precise ceramics have multiple physical and mechanical properties, such as high strength, high hardness, wear resistance, corrosion resistance, high-temperature resistance, and thermal shock resistance. Besides, precise ceramics also have great potential for usages in heat, light, sound, electricity, magnetism, chemistry, biology and other aspects.

To some degree, the performances of precise ceramics are far more efficient than those of modern high-quality alloys and polymer materials. Thus, precise ceramics play a leading role in the revolution of new materials. Meanwhile, precise ceramics have a wide range of applications in many industries, like petroleum, chemical, steel, electronics, textile, automobile industries, as well as aerospace, nuclear and military, and so on.

In recent decades, the application and development of ceramic materials have been very rapid. As one of the most potential developing materials after metal materials and polymer materials, ceramic materials have significantly superior comprehensive performance in various aspects of the metal materials and polymer materials currently used. In addition, studies have found that the wear rate of ceramics is different under different loads. The pore does not cause crack propagation at low load; in the case of high load, the pores become unstable, and cracks and expansion cracks will be formed in the pores, resulting in high wear rate of products and weak anti-wear mutation ability.

CERAMIC COMPOSITION

Ceramic composition and properties are the atomic and molecular nature of ceramic materials and their resulting characteristics and performance in industrial applications.

Industrial ceramics are commonly understood to be all industrially used materials that are inorganic, nonmetallic solids. Usually they are metal oxides (that is, compounds of metallic elements and oxygen), but many ceramics (especially advanced ceramics) are compounds of metallic elements and carbon, nitrogen, or sulfur. In atomic structure they are most often crystalline, although they also may contain a combination of glassy and crystalline phases. These structures and chemical ingredients, though various, result in universally recognized ceramic-like properties of enduring utility, including the following: mechanical strength in spite of brittleness; chemical durability against the deteriorating effects of oxygen, water, acids, bases, salts, and organic solvents; hardness, contributing to resistance against wear; thermal and electrical conductivity considerably lower than that of metals; and an ability to take a decorative finish.

The relation between the properties of ceramics and their chemical and structural nature is described. Before such a description is attempted, though, it must be pointed out that there are exceptions to several of the defining characteristics outlined above. In chemical composition, for instance, diamond and graphite, which are two different forms of carbon, are considered to be ceramics even though they are not composed of inorganic compounds. There also are exceptions to the stereotypical properties ascribed to ceramics. To return to the example of diamond, this material, though considered to be a ceramic, has a thermal conductivity higher than that of copper—a property the jeweler uses to differentiate between true diamond and simulants such as cubic zirconia (a single-crystal form of zirconium dioxide). Indeed, many ceramics are quite conductive electrically. For instance, a polycrystalline (many-grained) version of zirconia is used as an oxygen sensor in automobile engines owing to its ionic conductivity. Also, copper oxide-based ceramics have been shown to have superconducting properties. Even the well-known brittleness of ceramics has its exceptions. For example, certain composite ceramics that contain whiskers, fibres, or particulates that interfere with crack propagation display flaw tolerance and toughness rivaling that of metals.

Nevertheless, despite such exceptions, ceramics generally display the properties of hardness, refractoriness (high melting point), low conductivity, and brittleness. These properties are intimately related to certain types of chemical bonding and crystal structures found in the material. Chemical bonding and crystal structure are addressed in turn below.

Chemical Bonds

Underlying many of the properties found in ceramics are the strong primary bonds that hold the atoms together and form the ceramic material. These chemical bonds are of two types: they are either ionic in character, involving a transfer of bonding electrons from electropositive atoms (cations) to electronegative atoms (anions), or they are covalent in character, involving orbital sharing of electrons between the constituent atoms or ions. Covalent bonds are highly directional in nature, often dictating the types of crystal structure possible. Ionic bonds, on the other hand, are entirely nondirectional. This nondirectional nature allows for hard-sphere packing arrangements of the ions into a variety of crystal structures, with two limitations. The first limitation involves the relative size of the anions and the cations. Anions are usually larger and close-packed, as in the face-centred cubic (fcc) or hexagonal close-packed (hcp) crystal structures found in metals. Cations, on the other hand, are usually smaller, occupying interstices, or spaces, in the crystal lattice between the anions.

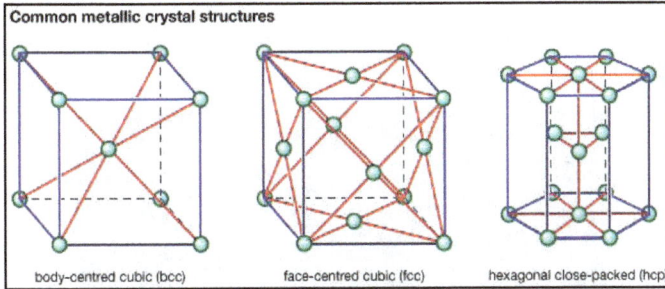

Figure: Three common metallic crystal structures.

The second limitation on the types of crystal structure that can be adopted by ionically bonded atoms is based on a law of physics—that the crystal must remain electrically neutral. This law of electroneutrality results in the formation of very specific stoichiometries—that is, specific ratios of cations to anions that maintain a net balance between positive and negative charge. In fact, anions are known to pack around cations, and cations around anions, in order to eliminate local charge imbalance. This phenomenon is referred to as coordination.

Most of the primary chemical bonds found in ceramic materials are actually a mixture of ionic and covalent types. The larger the electronegativity difference between anion and cation (that is, the greater the difference in potential to accept or donate electrons), the more nearly ionic is the bonding (that is, the more likely are electrons to be transferred, forming positively charged cations and negatively charged anions). Conversely, small differences in electronegativity lead to a sharing of electrons, as found in covalent bonds.

Secondary bonds also are important in certain ceramics. For example, in diamond, a single-crystal form of carbon, all bonds are primary, but in graphite, a polycrystalline form of carbon, there are primary bonds within sheets of crystal grains and secondary bonds between the sheets. The relatively weak secondary bonds allow the sheets to slide past one another, giving graphite the lubricity for which it is well known. It is the primary bonds in ceramics that make them among the strongest, hardest, and most refractory materials known.

Crystal Structure

Crystal structure is also responsible for many of the properties of ceramics. Each collection of ions is shown in an overall box that describes the unit cell of that structure. By repeatedly translating the unit cell one box in any direction and by repeatedly depositing the pattern of ions within that cell at each new position, any size crystal can be built up. In the first structure the material shown is magnesia (MgO), though the structure itself is referred to as rock salt because common table salt (sodium chloride, NaCl) has the same structure. In the rock salt structure each ion is surrounded by six immediate neighbours of the opposite charge (e.g., the central Mg^{2+} cation, which is surrounded by O^{2-} anions). This extremely efficient packing allows for local neutralization of charge and makes for stable bonding. Oxides that crystallize in this structure tend to have relatively high melting points. (Magnesia, for example, is a common constituent in refractory ceramics).

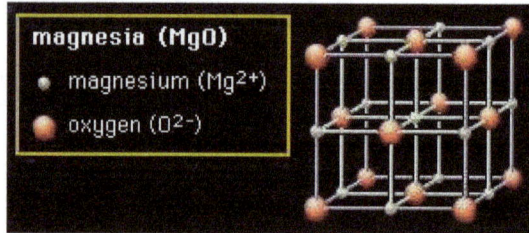

Figure: The arrangement of magnesium and oxygen ions in magnesia (MgO); an example of therock salt crystal structure.

The second is called fluorite, after the mineral calcium fluoride (CaF_2), which possesses this structure—though the material shown is urania (uranium dioxide, UO_2). In this structure the oxygen anions are bonded to only four cations. Oxides with this structure are well known for the ease with which oxygen vacancies can be formed. In zirconia (zirconium dioxide, ZrO_2), which also possesses this structure, a great number of vacancies can be formed by doping, or carefully inserting ions of a different element into the composition. These vacancies become mobile at high temperatures, imparting oxygen-ion conductivity to the material and making it useful in certain electrical applications. The fluorite structure also exhibits considerable open space, especially at the centre of the unit cell. In urania, which is used as a fuel element in nuclear reactors, this openness is believed to help accommodate fission products and reduce unwanted swelling.

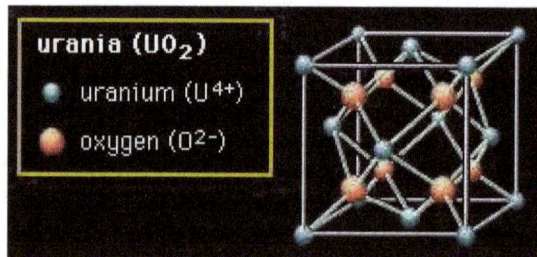

Figure: The arrangement of uranium and oxygen ions in urania (UO_2); an example of the fluorite crystal structure.

The third structure is called perovskite. In most cases the perovskite structure is cubic—that is, all sides of the unit cell are the same. However, in barium titanate ($BaTiO_3$), shown in the figure, the central Ti^{4+} cation can be induced to move off-centre, leading to a noncubic symmetry and to an electrostatic dipole, or alignment of positive and negative charges toward opposite ends of the structure. This dipole is responsible for the ferroelectric properties of barium titanate, in which domains of neighbouring dipoles line up in the same direction. The enormous dielectric constants achievable with perovskite materials are the basis of many ceramic capacitor devices.

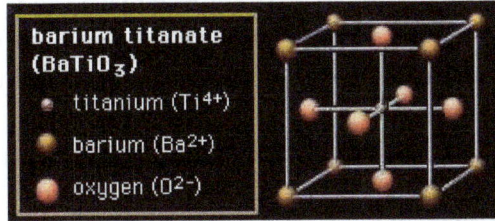

Figure: The arrangement of titanium, barium, and oxygen ions in
barium titanate (BaTiO$_3$); an example of the perovskite
crystal structure.

The noncubic variations found in perovskite ceramics introduce the concept of anisotropy—i.e., an ionic arrangement that is not identical in all directions. In severely anisotropic materials there can be great variation of properties. These cases are illustrated by yttrium barium copper oxide (YBCO; chemical formula YBa$_2$Cu$_3$O$_7$), shown in figure. YBCO is a superconducting ceramic; that is, it loses all resistance to electric current at extremely low temperatures. Its structure consists of three cubes, with yttrium or barium at the centre, copper at the corners, and oxygen at the middle of each edge—with the exception of the middle cube, which has oxygen vacancies at the outer edges. The critical feature in this structure is the presence of two sheets of copper-oxygen ions, located above and below the oxygen vacancies, along which superconduction takes place. The transport of electrons perpendicular to these sheets is not favoured, making the YBCO structure severely anisotropic. (One of the challenges in fabricating crystalline YBCO ceramics capable of passing large currents is to align all the grains in such a manner that their copper-oxygen sheets line up).

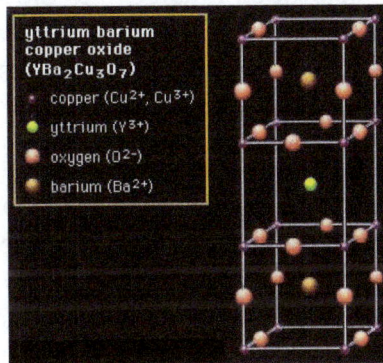

Figure: The arrangement of copper, yttrium, oxygen, and barium
ions in yttrium barium copper oxide (YBa$_2$Cu$_3$O$_7$); an example of a
superconducting ceramic crystal structure.

Nonconductivity

Ordinarily, ceramics are poor conductors of electricity and therefore make excellent insulators. Nonconductivity arises from the lack of "free" electrons such as those found in metals. In ionically bonded ceramics, bonding electrons are accepted by the electro-negative elements, such as oxygen, and donated by the electropositive elements, usual-ly a metal. The result is that all electrons are tightly bound to the ions in the structure, leaving no free electrons to conduct electricity. In covalent bonding, bonding electrons are similarly localized in the directional orbitals between the atoms, and there are no free electrons to conduct electricity.

There are two ways that ceramics can be made electrically conductive. At sufficiently high temperatures point defects such as oxygen vacancies can arise, leading to ionic conductivity. (This is pointed out in the case of zirconia). In addition, the introduction of certain transition-metal elements (such as iron, copper, manganese, or cobalt), lan-thanoid elements (such as cerium), or actinoid elements (such as uranium) can pro-duce special electronic states in which mobile electrons or electron holes arise. The copper-based superconductors are a good example of conductive transition-metal ox-ide ceramics—in this case, conductivity arising at extremely low temperatures.

Brittleness

Unlike most metals, nearly all ceramics are brittle at room temperature; i.e., when subjected to tension, they fail suddenly, with little or no plastic deformation prior to fracture. Metals, on the other hand, are ductile (that is, they deform and bend when subjected to stress), and they possess this extremely useful property owing to imperfec-tions called dislocations within their crystal lattices. There are many kinds of disloca-tions. In one kind, known as an edge dislocation, an extra plane of atoms can be gener-ated in a crystal structure, straining to the breaking point the bonds that hold the atoms together. If stress were applied to this structure, it might shear along a plane where the bonds were weakest, and the dislocation might slip to the next atomic position, where the bonds would be re-established. This slipping to a new position is at the heart of plastic deformation. Metals are usually ductile because dislocations are common and are normally easy to move.

In ceramics, however, dislocations are not common (though they are not nonexistent), and they are difficult to move to a new position. The reasons for this lie in the nature of the bonds holding the crystal structure together. In ionically bonded ceramics some planes—such as the so-called plane shown slicing diagonally through the rock salt struc-ture in figure, top—contain only one kind of ion and are therefore unbalanced in their distribution of charges. Attempting to insert such a half plane into a ceramic would not favour a stable bond unless a half plane of the oppositely charged ion was also inserted. Even in the case of planes that were charge-balanced—for instance, the plane created by a vertical slice down the middle of the rock salt crystal structure, as shown in Fig-ure, bottom—slip induced along the middle would bring identically charged ions into proximity. The identical charges would repel each other, and dislocation motion would be impeded. Instead, the material would tend to fracture in the manner commonly as-sociated with brittleness.

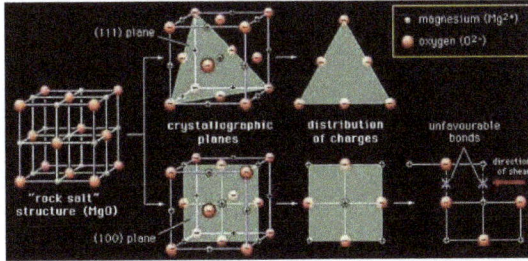

Figure: Barriers to slip in ceramic crystal structures.

Beginning with the rock salt structure of magnesia (MgO; shown at left), in which there is a stable balance of positive and negative charges, two possible crystallographic planes show the difficulty of establishing stable imperfections. The plane (shown at top) would contain atoms of identical charge; inserted as an imperfection into the crystal structure, such an imbalanced distribution of charges would not be able to establish a stable bond. The plane would show a balance between positive and negative charges, but a shear stress applied along the middle of the plane would force identically charged atoms into proximity—again creating a condition unfavourable for stable bonding.

In order for polycrystalline materials to be ductile, they must possess more than a minimum number of independent slip systems—that is, planes or directions along which slip can occur. The presence of slip systems allows the transfer of crystal deformations from one grain to the next. Metals typically have the required number of slip systems, even at room temperature. Ceramics, however, do not, and as a result they are notoriously brittle.

Glasses, which lack a long-range periodic crystal structure altogether, are even more susceptible to brittle fracture than ceramics. Because of their similar physical attributes (including brittleness) and similar chemical constituents (e.g., oxides), inorganic glasses are considered to be ceramics in many countries of the world. Indeed, partial melting during the processing of many ceramics results in a significant glassy portion in the final makeup of many ceramic bodies (for instance, porcelains), and this portion is responsible for many desirable properties (e.g., liquid impermeability).

Powder Processing

Unlike metals and glasses, which can be cast from the melt and subsequently rolled, drawn, or pressed into shape, ceramics must be made from powders. As pointed out above, ceramics are seldom deformable, especially at room temperature, and the microstructural modifications achieved by cold-working and recrystallizing metals are impossible with most ceramics. Instead, ceramics are usually made from powders, which are consolidated and densified by sintering. Sintering is a process whereby particles bond and coalesce under the influence of heat, leading to shrinkage and reduction in porosity. A similar process in metal manufacturing is referred to as powder metallurgy.

Powder processing is used to make products that are normally identified as traditional ceramics—namely, whitewares such as porcelain and china, structural clay products such as brick and tile, refractories for insulating and lining metallurgical furnaces and glass tanks, abrasives, and cements. It also is used in the production of advanced ceramics, including ceramics for electronic, magnetic, optical, nuclear, and biological applications. Traditional ceramics involve large volumes of product and relatively low value-added manufacturing. Advanced ceramics, on the other hand, tend to involve smaller volumes of product and higher value-added manufacturing.

CERAMIC GLAZE

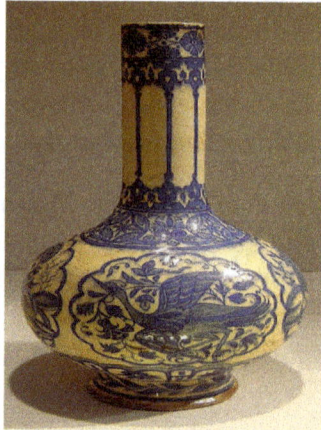

Composite body, painted, and glazed bottle.

Ceramic glaze is an impervious layer or coating of a vitreous substance which has been fused to a ceramic body through firing. Glaze can serve to color, decorate or waterproof an item. Glazing renders earthenware vessels suitable for holding liquids, sealing the inherent porosity of unglazed biscuit earthenware. It also gives a tougher surface. Glaze is also used on stoneware and porcelain. In addition to their functionality, glazes can form a variety of surface finishes, including degrees of glossy or matte finish and color. Glazes may also enhance the underlying design or texture either unmodified or inscribed, carved or painted.

Detail of dripping rice-straw ash glaze (top).

Most pottery produced in recent centuries has been glazed, other than pieces in unglazed biscuit porcelain, terracotta, or some other types. Tiles are almost always glazed on the surface face, and modern architectural terracotta is very often glazed. Glazed brick is also common. Domestic sanitary ware is invariably glazed, as are many ceramics used in industry, for example ceramic insulators for overhead power lines.

The most important groups of traditional glazes, each named after its main ceramic fluxing agent, are:

- Ash glaze, important in East Asia, simply made from wood or plant ash, which contains potash and lime.

- Feldspathic glazes of porcelain.

- Lead glazes, plain or coloured, are shiny and transparent after firing, which need only about 800 °C (1,470 °F). They have been used for about 2,000 years around the Mediterranean, in Europe, and China. They includes *sancai* and Victorian majolica.

- Salt-glaze, mostly European stoneware. It uses ordinary salt.

- Tin-glaze, which coats the ware with lead glaze made opaque white by the addition of tin. Known in the Ancient Near East and then important in Islamic pottery, from which it passed to Europe. Includes Hispano-Moresque ware, maiolica (also called majolica), faience, and Delftware.

Modern materials technology has invented new vitreous glazes that do not fall into these traditional categories.

Composition

Glazes need to include a ceramic flux which functions by promoting partial liquefaction in the clay bodies and the other glaze materials. Fluxes lower the high melting point of the glass formers silica, and sometimes boron trioxide. These glass formers may be included in the glaze materials, or may be drawn from the clay beneath.

Raw materials of ceramic glazes generally include silica, which will be the main glass former. Various metal oxides, such as sodium, potassium, and calcium, act as flux and therefore lower the melting temperature. Alumina, often derived from clay, stiffens the molten glaze to prevent it from running off the piece. Colorants, such as iron oxide, copper carbonate, or cobalt carbonate, and sometimes opacifiers like tin oxide or zirconium oxide, are used to modify the visual appearance of the fired glaze.

Process

Glaze may be applied by dry-dusting a dry mixture over the surface of the clay body or by inserting salt or soda into the kiln at high temperatures to create an atmosphere rich in sodium vapor that interacts with the aluminium and silica oxides in the body to form and deposit glass, producing what is known as salt glaze pottery. Most commonly, glazes in aqueous suspension of various powdered minerals and metal oxides are applied by dipping pieces directly into the glaze. Other techniques include pouring the glaze over the piece, spraying it onto the piece with an airbrush or similar tool, or applying it directly with a brush or other tool.

Iznik tiles in the Enderûn Library, Topkapi Palace.

To prevent the glazed article from sticking to the kiln during firing, either a small part of the item is left unglazed, or it's supported on small refractory supports such as kiln spurs and Stilts that are removed and discarded after the firing. Small marks left by these spurs are sometimes visible on finished ware.

Decoration applied under the glaze on pottery is generally referred to as underglaze. Underglazes are applied to the surface of the pottery, which can be either raw, "greenware", or "biscuit"-fired (an initial firing of some articles before the glazing and re-firing). A wet glaze—usually transparent—is applied over the decoration. The pigment fuses with the glaze, and appears to be underneath a layer of clear glaze. An example of underglaze decoration is the well-known "blue and white" porcelain famously produced in Germany, England, the Netherlands, China, and Japan. The striking blue color uses cobalt as cobalt oxide or cobalt carbonate.

Sancai lead-glazed figure of heavenly
guardian, Tang dynasty.

Decoration applied on top of a layer of glaze is referred to as overglaze. Overglaze methods include applying one or more layers or coats of glaze on a piece of pottery or by applying a non-glaze substance such as enamel or metals (e.g., gold leaf) over the glaze.

Overglaze colors are low-temperature glazes that give ceramics a more decorative, glassy look. A piece is fired first, this initial firing being called the glost firing, then the overglaze decoration is applied, and it is fired again. Once the piece is fired and comes out of the kiln, its texture is smoother due to the glaze.

Environmental Impact

As of 2012, over 650 ceramic manufacturing establishments were reported in the United States, with likely many more across the developed and developing world. Floor tile, wall tile, sanitary-ware, bathroom accessories, kitchenware, and tableware are all potential ceramic-containing products that are available for consumers. Heavy metals are dense metals used in glazes to produce a particular color or texture. Glaze components are more likely to be leached into the environment when non-recycled ceramic products are exposed to warm or acidic water. Leaching of heavy metals occurs when ceramic products are glazed incorrectly or damaged. Lead and chromium are two heavy metals commonly used in ceramic glazes that are heavily monitored by government agencies due to their toxicity and ability to bioaccumulate.

Metal Oxide Chemistry

Metals used in ceramic glazes are typically in the form of metal oxides.

Lead(II) Oxide

Ceramic manufacturers primarily use lead(II) oxide (PbO) as a flux for its low melting range, wide firing range, low surface tension, high index of refraction, and resistance to devitrification.

In polluted environments, nitrogen dioxide reacts with water (H_2O) to produce nitrous acid (HNO_2) and nitric acid (HNO_3).

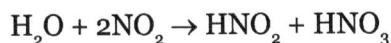

$$H_2O + 2NO_2 \rightarrow HNO_2 + HNO_3$$

Soluble Lead(II) nitrate ($Pb(NO_3)_2$) forms when lead(II) oxide (PbO) of leaded glazes is exposed to nitric acid (HNO_3).

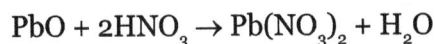

$$PbO + 2HNO_3 \rightarrow Pb(NO_3)_2 + H_2O$$

Because lead exposure is strongly linked to a variety of health problems, collectively referred to as lead poisoning, the disposal of leaded glass (chiefly in the form of discarded CRT displays) and lead-glazed ceramics is subject to toxic waste regulations.

Chromium(III) Oxide

Chromium(III) oxide (Cr_2O_3) is used as a colorant in ceramic glazes. Chromium(III) oxide can undergo a reaction with calcium oxide (CaO) and atmospheric oxygen in temperatures reached by a kiln to produce calcium chromate ($CaCrO_4$). The oxidation reaction changes chromium from its +3 oxidation state to its +6 oxidation state. Chromium(VI) is very soluble and the most mobile out of all the other stable forms of chromium.

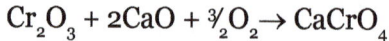

$$Cr_2O_3 + 2CaO + \tfrac{3}{2}O_2 \rightarrow CaCrO_4$$

Chromium may enter water systems via industrial discharge. Chromium(VI) can enter the environment directly or oxidants present in soils can react with chromium(III) to produce chromium(VI). Plants have reduced amounts of chlorophyll when grown in the presence of chromium(VI).

Prevention

Chromium oxidation during manufacturing processes can be reduced with the introduction of compounds that bind to calcium. Ceramic industries are reluctant to use lead alternatives since leaded glazes provide products with a brilliant shine and smooth surface. The United States Environmental Protection Agency has experimented with a dual glaze, barium alternative to lead, but they were unsuccessful in achieving the same optical effect as leaded glazes.

Meissen porcelain, with blue underglaze decoration on porcelain.

Coloured lead glazes majolica circa 1870.

TRIBOLOGY OF CERAMICS

Characteristics of friction and wear of a ceramic material are determined by a combination of its bulk microstructure parameters, surface conditions and environmental factors (temperature, atmosphere pressure, etc).

Effect of Microstructure on Tribological Properties of Ceramics

Parameters of microstructure and their influence on friction and wear of ceramics:

- Grain size:

  The main disadvantage of Ceramics as compared to Metals and Polymers is their low fracture toughness.

  Toughness is a bulk mechanical property of a material however it correlates with its wear resistance particularly when the wear is a result of abrasive action caused by cracking.

  Finer grain structure results in increased toughness and better wear resistance.

  Grain size also determines the surface finish quality, which may be achieved by grinding and polishing operations.

  Fine grain structure allows to decrease the size of the surface microasperities after the surface finish operation resulting in lower coefficient of friction.

- Critical flaw size (the size of a flaw that results in rapid fracture):

  Effect of flaw size on the fracture strength of a ceramic material is expressed by the Griffith equation:

  $$\sigma_C = K_{IC} / (Y(\pi a)^{1/2})$$

  where:

  K_{IC} – Stress-intensity factor, measured in $MPa*m^{1/2}$;

  a – The flaw size;

  Y – Geometry factor.

According to the equation flaws of lower size result in increased material toughness and higher wear resistance. Flaw size is generally proportional to the grain size.

- Homogeneity:

  Homogeneous distribution of the matrix particles size and pores size, second phase particles (toughening particles) incorporated between the matrix particles, aid phase (binders, etc.) locating at the grains boundaries results in lowering the flaw size and consequently in increase of the fracture strength (according to the Griffith equation).

  Higher fracture strength causes higher wear resistance. Bulk homogeneity of the microstructure allows creating fine and homogeneous surface finish with low content of surface flaws. High quality surface possess low coefficient of friction.

Manufacturing Processes Forming Microstructure of Ceramics

- Powder preparation:

  Powder characteristics such as particle shape (spherical, irregular), average particle size, size distribution determine the ceramic grain size and the amount and size of the pores.

- Compaction (shape forming):

  The value of the applied pressure, the method of its application (Uniaxial (Die) Pressing, Isostatic Pressing, Injection Molding, Extrusion, Slip Casting, etc.) and the amount of binders and other additives (plasticizers, lubricants, deflocculants, water etc.) determine the pores size and the residual internal stresses.

- Sintering:

  Diffusion proceeding during sintering process causes the pores to diminish or even to close up resulting in densification of the ceramic material.

  The bonding and other second phases are distributed between the grains of the main ceramic phase.

  The matrix grains may grow during the sintering process. Thus sintering process determines the final grains and pores size and the physical and chemical homogeneity.

Tribological Properties of Ceramics

Surface Characteristics

- Surface Topography:

  Friction characteristics (coefficient of friction, wear) are strongly dependent on the type of the lubrication regime (boundary lubrication, mixed lubrication, hydrodynamic lubrication).

  The lubrication regime is determined by the ratio of the lubricant film thickness to the surface roughness Ra.

  Rough ceramic surface with relatively large microasperities causes direct contact between the rubbing surfaces and results in high coefficient of friction and increased wear.

  High surface finish quality allows to improve the tribological characteristics of ceramics.

  Ceramics are brittle and they wear by fracture mechanism, which is characterized by formation of cracks in the subsurface regions surrounding the wear groove. The volume of the lost material is higher than the volume of the wear track.

Thus wear of brittle ceramics results in roughening the surface. The effect of roughening during friction is lower in toughened ceramics.

- Surface defects:

Sintering defects, surface machining, impacts during friction, embedded particles introduce surface flaws, which lead to fracture cracking and increase wear.

- Surface composition and tribochemical reactions:

Ceramic surface may adsorb molecules of the environmental gases and liquids. Such surfaces with modified composition may have different coefficient of friction.

Coefficient of friction of ceramics in vacuum is commonly higher than that in air.

Hydration of Oxide ceramics in a humid atmosphere also results in changing their coefficients of friction and wear. Wear of hydrated silicon nitride and silicon carbide is decreased. Wear of hydrated Alumina ceramics and Zirconia ceramics is increased due to chemisorption embrittlement.

Surface of Non-oxide ceramics oxidizes in the presence of Oxygen in the environment. The oxidation is enhanced at increased temperatures.

Oxide film on the surface of a non-oxide ceramic decreases the coefficients of friction serving as a solid lubricant.

Methods of Modification of Ceramic Surfaces

- Plasma oxidizing: A method of surface oxidation by elemental Oxygen supplied to the ceramic surface by plasma.

- Ion nitriding and carburizing: A method of introducing nitrogen (nitriding) or carbon (carburizing) atoms into the ceramic surface by means of plasma (glow-discharge).

- Ion implantation: A method of introducing a material into a ceramic surface by electrostatically accelerated ions.

- Laser densification: A method of heating the ceramic surface layer by a laser beam resulting in closing the pores between the ceramic powder particles.

- Electron beam densification: A method of heating the ceramic surface layer by an electron beam resulting in closing the pores between the ceramic powder particles.

- Chemical etching: Cleaning the ceramic surface by acids.

- Sputter etching: Bombarding the ceramic surface by accelerated plasma ions, which vaporize the surface molecules.

Effect of Lubrication on Tribological Properties of Ceramics

Lubricants decrease coefficient of friction and reduce wear of the rubbing parts.

Lubricants remove the heat generated by friction. This function is particularly important for ceramics since they have lower thermal conductivity and usually produce more heat due to relatively high coefficient of friction.

Lubricants remove wear debris from the rubbing surfaces.

Lubricants also protect the ceramic surface from the environment.

- Liquid lubricants:

  Liquid hydrocarbon lubricants are commonly used for relatively low temperatures (up to 392 °F/200 °C). Silicone oils may be used up to 570 °F (300 °C).

- Solid lubricants:

  Solid lubricants may be used for lubricating ceramics in various forms: Suspensions in liquid lubricants, dry powders, Dispersions in gases, coatings.

  Requirements to solid lubricants properties: good adhesion to the ceramic surface, low shear strength in the sliding direction and high compression strength in the direction of the load (perpendicular to the sliding direction).

  Substances used as solid lubricants: graphite, molybdenum disulfide, boron nitride, Polytetrafluoroethylene (PTFE), calcium fluoride-barium fluoride eutectic.

  Maximum work temperature some of the solid lubricants is low (PTFE: 392 °F/200 °C). Other lubricants may withstand up to 1508 °F/820 °C (calcium fluoride-barium fluoride eutectic).

- Gaseous lubricants:

  Vapors of some organic substances may serve as lubricants for ceramics. The vaporized molecules of such lubricant reach the ceramic surface react with it and form on its surface a film possessing low coefficient of friction.

References

- What-are-ceramic-materials-and-their-uses, manufacturing-technology: brighthubengineering. com, Retrieved 2 June, 2019

- Properties: dynacer.com, Retrieved 22 May, 2019

- Advanced-ceramics, technology: britannica.com, Retrieved 13 March, 2019

- Super-hard-ceramics: precision-ceramics.co.uk, Retrieved 29 June, 2019

- Ceramic-composition-and-properties-103137: britannica.com, Retrieved 14 July, 2019

- Division, Company Statistics. "Statistics of U.S. Businesses Main Page". Www.census.gov. Archived from the original on 2015-11-26. Retrieved 2015-11-27

- Traditional-ceramics, technology: britannica.com, Retrieved 23 February, 2019

- Madan, Gaurav (2005). S.Chands Success Guide (Q&A) Inorganic Chemistry. S. Chand Publishing. ISBN 9788121918572

- Precise-ceramics-vs-traditional-ceramics: preciseceramic.com, Retrieved 31 March, 2019

- Daiheng, Gao (2002). Chinese Architecture – The Lia, Song, Xi Xia and Jin Dynasties (English ed.). Yale University Press. Pp. 166, 183. ISBN 978-0-300-09559-3

2

Classification of Ceramics

Ceramics are classified into various categories such as clay products, glass-ceramics, refractories, porcelain, ceramic-impregnated fabrics, terracotta, optical ceramics, abrasives, ceramics cements, and transparent ceramics. This chapter closely examines these categories of ceramics to provide an extensive understanding of the subject.

CLAY PRODUCTS

Clay products are one of the most important classes of structural materials. The raw materials used in their manufacture are clay blended with quartz, sand, chamatte (refractory clay burned at 1000-1400 °C and crushed), slag, sawdust and pulverized coal. Structural clay products or building ceramics are basically fabricated by moulding, drying and burning a clay mass. Higher the bulk specific gravity, the stronger is the clay product. This rule does not hold good for vitrified products since the specific gravity of clay decreases as vitrification advances.

Bulk specific gravity of clay brick ranges from 1.6 to 2.5.

According to the method of manufacture and structure, bricks, tiles, pipes, terracotta, earthenwares, stonewares, porcelain, and majolica are well recognized and employed in building construction. Clay bricks have pleasing appearance, strength and durability whereas clay tiles used for light-weight partition walls and floors possess high strength and resistance to fire. Clay pipes on account of their durability, strength, lightness and cheapness are successfully used in sewers, drains and conduits.

Polycrystalline materials and products formed by baking natural clays and mineral admixtures at a high temperature and also by sintering the oxides of various metals and other high melting-point inorganic substances.

Clay is the most important raw material used for making bricks. It is an earthen mineral mass or fragmentary rock capable of mixing with water and forming a plastic viscous mass which has a property of retaining its shape when moulded and dried. When such masses are heated to redness, they acquire hardness and strength. This is a result of micro-structural changes in clay and as such is a chemical property. Purest clays consist mainly of kaolinite ($2SiO_2.Al_2O_3.2H_2O$) with small quantities of minerals such as quartz, mica, felspar, calcite, magnesite, etc. By their origin, clays are subdivided as residual and transported clays. Residual clays, known as Kaolin or China clay, are formed from the decay of underlying rocks and are used for making pottery. The transported or sedimentary clays result from the action of weathering agencies. These are more disperse, contain impurities, and free from large particles of mother rocks.

On the basis of resistance to high temperatures (more than 1580 °C), clays are classified as refractory, high melting and low melting clays. The refractory clays are highly disperse and very plastic. These have high content of alumina and low content of impurities, such as Fe_2O_3, tending to lower the refractoriness. High melting clays have high refractoriness (1350-1580 °C) and contain small amount of impurities such as quartz, felspar, mica, calcium carbonate and magnesium carbonate. These are used for manufacturing facing bricks, floor tiles, sewer pipes, etc. Low melting clays have refractoriness less than 1350 °C and have varying compositions. These are used to manufacture bricks, blocks, tiles, etc.

Admixtures are added to clay to improve its properties, if desired. Highly plastic clays which require mixing water up to 28 per cent, give high drying and burning shrinkage, call for addition of lean admixtures or non-plastic substances such as quartz sand, chamottee, ash, etc. Items of lower bulk density and high porosity are obtained by addition of admixture that burn out. The examples of burning out admixtures are sawdust, coal fines, pulverized coal. etc. Acid resistance items and facing tiles are manufactured from clay by addition of water-glass or alkalis.

Burning temperature of clay items can be reduced by blending clay with fluxes such as felspar, iron bearing ores, etc. Plasticity of moulding mass may be increased by adding surfactants such as sulphite-sodium vinasse (0.1-0.3%).

Physical Properties of Clays

Plasticity, tensile strength, texture, shrinkage, porosity, fusibility and colour after burning are the physical properties which are the most important in determining the value of clay. Knowledge of these properties is of more benefit in judging the quality of the raw material than a chemical analysis.

By plasticity is meant the property which wetted clay has of being permanently deformed without cracking. The amount of water required by different clays to produce the most plastic condition varies from 15 to 35 per cent. Although plasticity is the most important physical property of clay, yet there are no methods of measuring it which are entirely satisfactory. The simplest and the most used test is afforded by feeling of the wetted clay with the fingers. Personal equation necessarily plays a large part in such determination.

Since clay ware is subjected to considerable stress in moulding, handling and drying, a high tensile strength is desirable. The test is made by determining the stregth of specimens which have been moulded into briquette form and very carefully dried.

The texture of clay is measured by the fineness of its grains. In rough work the per cent passing a No. 100 sieve is determined. No numerical limit to the grain size or desired relation between sizes has been established. Very fine grained clays free from sand are more plastic and shrink more than those containing coarser material.

Knowledge of shrinkage both in drying and in burning is required in order to produce a product of required size. Also the amount of shrinkage forms an index of the degree of burning. The shrinkage in drying is dependent upon pore space within the clay and upon the amount of mixing water. The addition of sand or ground burnt clay lowers shrinkage, increases the porosity and facilitates drying. Fire shrinkage is dependent upon the proportion of volatile elements, upon texture and the way that clay burns.

By porosity of clay is meant the ratio if the volume of pore space to the dry volume. Since porosity affects the proportion of water required to make clay plastic, it will indirectly influence air shrinkage. Large pores allow the water to evaporate more easily and consequently permit a higher rate of drying than do small pores. In as much as the rate at which the clay may be safely dried is of great importance in manufacturing clay products, the effect of porosity on the rate of drying should be considered.

The temperature at which clay fuses is determined by the proportion of fluxes, texture, homogeneity of the material, character of the flame and its mineral constitution. Owing to non-uniformity in composition, parts of the clay body melt at different rates so that the softening period extends over a considerable range both of time and temperature. This period is divided into incipient vitrification and viscous vitrification.

Experiments roughly indicate that the higher the proportion of fluxes the lower the melting point. Fine textured clays fuse more easily than those of coarser texture and the same mineral composition. The uniformity of the clay mass determines very largely the influence of various elements; the carbonate of lime in large lumps may cause popping when present in small percentages, but when finely ground 15 per cent of it may be allowed in making brick or tile. Lime combined with silicate of alumina (feldspar) forms a desirable flux. Iron in the ferrous form, found in carbonates and in magnetite, fuses more easily than when present as ferric iron. If the kiln atmosphere is insufficiently oxidizing in character during the early stages of burning, the removal of carbon and sulphur will be prevented until the mass has shrunk to such an extent as to prevent their expulsion and the oxidation of iron. When this happens, a product with a discoloured core or swollen body is likely to result.

A determination of the fusibility of a clay is of much importance both in judging of the cost of burning it and in estimating its refractoriness.

GLASS-CERAMIC

Glass-ceramics have an amorphous phase and one or more crystalline phases and are produced by a so-called "controlled crystallization" in contrast to a spontaneous crystallization, which is usually not wanted in glass manufacturing. Glass-ceramics have the fabrication advantage of glass, as well as special properties of ceramics. When used for sealing, some glass-ceramics do not require brazing but can withstand brazing temperatures up to 700 °C. Glass-ceramics usually have between 30% [m/m] and 90% [m/m] crystallinity and yield an array of materials with interesting properties like zero porosity, high strength, toughness, translucency or opacity, pigmentation, opalescence, low or even negative thermal expansion, high temperature stability, fluorescence, machinability, ferromagnetism, resorbability or high chemical durability, biocompatibility, bioactivity, ion conductivity, superconductivity, isolation capabilities, low dielectric constant and loss, high resistivity and break-down voltage. These properties can be tailored by controlling the base-glass composition and by controlled heat treatment/crystallization of base glass. In manufacturing, glass-ceramics are valued for having the strength of ceramic but the hermetic sealing properties of glass.

Glass-ceramics are mostly produced in two steps: First, a glass is formed by a glass-manufacturing process. The glass is cooled down and is then reheated in a second step. In this heat treatment the glass partly crystallizes. In most cases nucleation agents are added to the base composition of the glass-ceramic. These nucleation agents aid and control the crystallization process. Because there is usually no pressing and sintering, glass-ceramics have, unlike sintered ceramics, no pores.

A wide variety of glass-ceramic systems exists, e.g., the $Li_2O \times Al_2O_3 \times nSiO_2$ system (LAS system), the $MgO \times Al_2O_3 \times nSiO_2$ system (MAS system), the $ZnO \times Al_2O_3 \times nSiO_2$ system (ZAS system).

The LAS system mainly refers to a mix of lithium, silicon, and aluminum oxides with additional components, e.g., glass-phase-forming agents such as Na_2O, K_2O and CaO and refining agents. As nucleation agents most commonly zirconium(IV) oxide in combination with titanium(IV) oxide is used. This important system was studied first and intensively by Hummel, and Smoke.

After crystallization the dominant crystal phase in this type of glass-ceramic is a high-quartz solid solution (HQ s.s.). If the glass-ceramic is subjected to a more intense heat treatment, this HQ s.s. transforms into a keatite-solid solution (K s.s., sometimes wrongly named as beta-spodumene). This transition is non-reversible and reconstructive, which means bonds in the crystal-lattice are broken and new arranged. However, these two crystal phases show a very similar structure as Li could show.

The most interesting properties of these glass-ceramics are their thermomechanical properties. Glass-ceramic from the LAS system is a mechanically strong material and can sustain repeated and quick temperature changes up to 800–1000 °C. The dominant crystalline phase of the LAS glass-ceramics, HQ s.s., has a strong *negative* coefficient of thermal expansion (CTE), keatite-solid solution as still a negative CTE but much higher than HQ s.s. These negative CTEs of the crystalline phase contrasts with the positive CTE of the residual glass. Adjusting the proportion of these phases offers a wide range of possible CTEs in the finished composite. Mostly for today's applications a low or even zero CTE is desired. Also a negative CTE is possible, which means, in contrast to most materials when heated up, such a glass-ceramic contracts. At a certain point, generally between 60% [m/m] and 80% [m/m] crystallinity, the two coefficients balance such that the glass-ceramic as a whole has a thermal expansion coefficient that is very close to zero. Also, when an interface between material will be subject to thermal fatigue, glass-ceramics can be adjusted to match the coefficient of the material they will be bonded to.

Originally developed for use in the mirrors and mirror mounts of astronomical telescopes, LAS glass-ceramics have become known and entered the domestic market through its use in glass-ceramic cooktops, as well as cookware and bakeware or as high-performance reflectors for digital projectors.

Ceramic Matrix Composites

One particularly notable use of glass-ceramics is in the processing of ceramic matrix composites. For many ceramic matrix composites typical sintering temperatures and times cannot be used, as the degradation and corrosion of the constituent fibres becomes more of an issue as temperature and sintering time increase. One example of this is SiC fibres, which can start to degrade via pyrolysis at temperatures above 1470K. One solution to this is to use the glassy form of the ceramic as the sintering feedstock rather than the ceramic, as unlike the ceramic the glass pellets have a softening point and will generally flow at much lower pressures and temperatures. This allows the use of less extreme processing parameters, making the production of many new technologically important fibre-matrix combinations by sintering possible.

Cooktops

Glass-ceramic from the LAS-System is a mechanically strong material and can sustain repeated and quick temperature changes. However, it is not totally unbreakable. Because it is still a brittle material as glass and ceramics are, it can be broken. There have been instances where users reported damage to their cooktops when the surface was struck with a hard or blunt object (such as a can falling from above or other heavy items).

The material has a very low heat conduction coefficient, which means that it stays cool outside the cooking area. It can be made nearly transparent (15–20% loss in a typical cooktop) for radiation in the infrared wavelengths.

In the visible range glass-ceramics can be transparent, translucent or opaque and even colored by coloring agents.

A glass-ceramic cooktop.

Today, there are two major types of electrical stoves with cooktops made of glass-ceramic:

- A glass-ceramic stove uses radiant heating coils or infrared halogen lamps as the heating elements. The surface of the glass-ceramic cooktop above the burner heats up, but the adjacent surface remains cool because of the low heat conduction coefficient of the material.

- An induction stove heats a metal pot's bottom directly through electromagnetic induction.

This technology is not entirely new, as glass-ceramic ranges were first introduced in the 1970s using Corningware tops instead of the more durable material used today. These first generation smoothtops were problematic and could only be used with flat-bottomed cookware as the heating was primarily conductive rather than radiative.

Compared to conventional kitchen stoves, glass-ceramic cooktops are relatively simple to clean, due to their flat surface. However, glass-ceramic cooktops can be scratched very easily, so care must be taken not to slide the cooking pans over the surface. If food with a high sugar content (such as jam) spills, it should never be allowed to dry on the surface, otherwise damage will occur.

For best results and maximum heat transfer, all cookware should be flat-bottomed and matched to the same size as the burner zone.

SILICON NITRIDE

Silicon nitride is a chemical compound of the elements silicon and nitrogen. Si_3N_4 is the most thermodynamically stable of the silicon nitrides. Hence, Si_3N_4 is the most commercially important of the silicon nitrides and is generally understood as what is being referred to where the term "silicon nitride" is used. It is a white, high-melting-point solid that is relatively chemically inert, being attacked by dilute HF and hot H_2SO_4. It is very hard (8.5 on the mohs scale). It has a high thermal stability.

Production

The material is prepared by heating powdered silicon between 1300 °C and 1400 °C in a nitrogen environment:

$$3\,Si + 2\,N_2 \rightarrow Si_3N_4$$

The silicon sample weight increases progressively due to the chemical combination of silicon and nitrogen. Without an iron catalyst, the reaction is complete after several hours (~7), when no further weight increase due to nitrogen absorption (per gram of silicon) is detected. In addition to Si_3N_4, several other silicon nitride phases (with chemical formulas corresponding to varying degrees of nitridation/Si oxidation state) have been reported in the literature, for example, the gaseous disilicon mononitride (Si_2N); silicon mononitride (SiN), and silicon sesquinitride (Si_2N_3), each of which are stoichiometric phases. As with other refractories, the products obtained in these high-temperature syntheses depends on the reaction conditions (e.g. time, temperature, and starting materials including the reactants and container materials), as well as the mode of purification. However, the existence of the sesquinitride has since come into question.

It can also be prepared by diimide route:

$$SiCl_4 + 6\,NH_3 \rightarrow Si(NH)_2 + 4\,NH_4Cl(s) \text{ at } 0\ ^\circ C$$

$$3\,Si(NH)_2 \rightarrow Si_3N_4 + N_2 + 3\,H_2(g) \text{ at } 1000\ ^\circ C$$

Carbothermal reduction of silicon dioxide in nitrogen atmosphere at 1400–1450 °C has also been examined:

$$3\,SiO_2 + 6\,C + 2\,N_2 \rightarrow Si_3N_4 + 6\,CO$$

The nitridation of silicon powder was developed in the 1950s, following the "rediscovery" of silicon nitride and was the first large-scale method for powder production. However, use of low-purity raw silicon caused contamination of silicon nitride by silicates and iron. The diimide decomposition results in amorphous silicon nitride, which needs further annealing under nitrogen at 1400–1500 °C to convert it to crystalline powder; this is now the second-most important route for commercial production. The carbothermal reduction was the earliest used method for silicon nitride production and is now considered as the most-cost-effective industrial route to high-purity silicon nitride powder.

Electronic-grade silicon nitride films are formed using chemical vapor deposition (CVD), or one of its variants, such as plasma-enhanced chemical vapor deposition (PECVD):

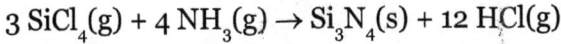

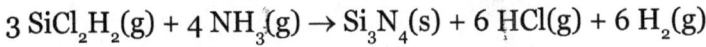

$$3\ SiH_4(g) + 4\ NH_3(g) \rightarrow Si_3N_4(s) + 12\ H_2(g) \text{ at } 750\text{-}850\ ^\circ C$$

$$3\ SiCl_4(g) + 4\ NH_3(g) \rightarrow Si_3N_4(s) + 12\ HCl(g)$$

$$3\ SiCl_2H_2(g) + 4\ NH_3(g) \rightarrow Si_3N_4(s) + 6\ HCl(g) + 6\ H_2(g)$$

For deposition of silicon nitride layers on semiconductor (usually silicon) substrates, two methods are used:

1. Low pressure chemical vapor deposition (LPCVD) technology, which works at rather high temperature and is done either in a vertical or in a horizontal tube furnace.

2. Plasma-enhanced chemical vapor deposition (PECVD) technology, which works at rather low temperature and vacuum conditions.

The lattice constants of silicon nitride and silicon are different. Therefore, tension or stress can occur, depending on the deposition process. Especially when using PECVD technology this tension can be reduced by adjusting deposition parameters.

Silicon nitride nanowires can also be produced by sol-gel method using carbothermal reduction followed by nitridation of silica gel, which contains ultrafine carbon particles. The particles can be produced by decomposition of dextrose in the temperature range 1200–1350 °C. The possible synthesis reactions are:

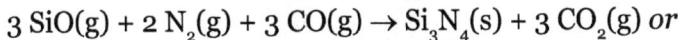

$$SiO_2(s) + C(s) \rightarrow SiO(g) + CO(g)\ and$$

$$3\ SiO(g) + 2\ N_2(g) + 3\ CO(g) \rightarrow Si_3N_4(s) + 3\ CO_2(g)\ or$$

$$3\ SiO(g) + 2\ N_2(g) + 3\ C(s) \rightarrow Si_3N_4(s) + 3\ CO(g).$$

Processing

Silicon nitride is difficult to produce as a bulk material—it cannot be heated over 1850 °C, which is well below its melting point, due to dissociation to silicon and nitrogen. Therefore, application of conventional hot press sintering techniques is problematic. Bonding of silicon nitride powders can be achieved at lower temperatures through adding additional materials (sintering aids or "binders") which commonly induce a degree of liquid phase sintering. A cleaner alternative is to use spark plasma sintering where heating is conducted very rapidly (seconds) by passing pulses of electric current through the compacted powder. Dense silicon nitride compacts have been obtained by this techniques at temperatures 1500–1700 °C.

Crystal Structure and Properties

Blue atoms are nitrogen and grey are silicon atoms:

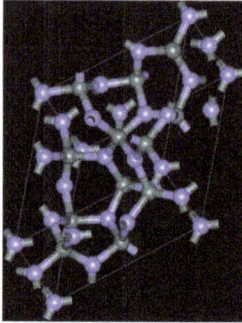 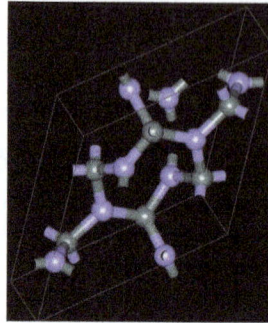 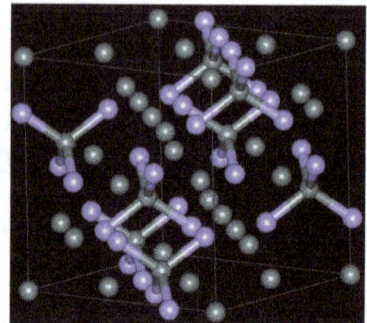

trigonal α-Si₃N₄.   hexagonal β-Si₃N₄   cubic γ-Si₃N₄

There exist three crystallographic structures of silicon nitride (Si_3N_4), designated as α, β and γ phases. The α and β phases are the most common forms of Si_3N_4, and can be produced under normal pressure condition. The γ phase can only be synthesized under high pressures and temperatures and has a hardness of 35 GPa.

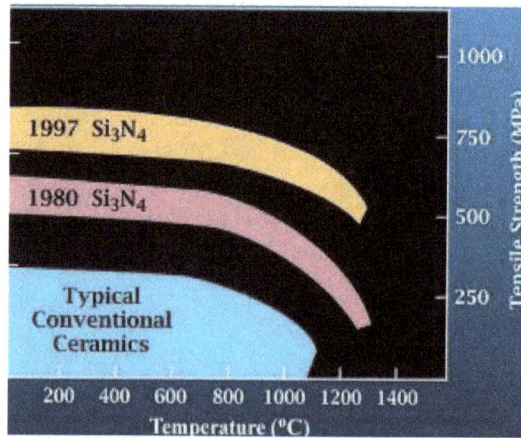

The α- and β-Si_3N_4 have trigonal and hexagonal structures, respectively, which are built up by corner-sharing SiN_4 tetrahedra. They can be regarded as consisting of layers of silicon and nitrogen atoms in the sequence ABAB... or ABCDABCD... in β-Si_3N_4 and α-Si_3N_4, respectively. The AB layer is the same in the α and β phases, and the CD layer in the α phase is related to AB by a c-glide plane. The Si_3N_4 tetrahedra in β-Si_3N_4 are interconnected in such a way that tunnels are formed, running parallel with the c axis of the unit cell. Due to the c-glide plane that relates AB to CD, the α structure contains cavities instead of tunnels. The cubic γ-Si_3N_4 is often designated as c modification in the literature, in analogy with the cubic modification of boron nitride (c-BN). It has a spinel-type structure in which two silicon atoms each coordinate six nitrogen atoms octahedrally, and one silicon atom coordinates four nitrogen atoms tetrahedrally.

The longer stacking sequence results in the α-phase having higher hardness than the β-phase. However, the α-phase is chemically unstable compared with the β-phase. At high temperatures when a liquid phase is present, the α-phase always transforms into the β-phase. Therefore, β-Si_3N_4 is the major form used in Si_3N_4 ceramics.

In addition to the crystalline polymorphs of silicon nitride, glassy amorphous materials may be formed as the pyrolysis products of preceramic polymers, most often containing varying amounts of residual carbon (hence they are more appropriately considered as silicon carbonitrides). Specifically, polycarbosilazane can be readily converted to an amorphous form of silicon carbonitride based material upon pyrolysis, with valuable implications in the processing of silicon nitride materials through processing techniques more commonly used for polymers.

Applications

In general, the main issue with applications of silicon nitride has not been technical performance, but cost. As the cost has come down, the number of production applications is accelerating.

Automobile Industry

One of the major applications of sintered silicon nitride is in automobile industry as a material for engine parts. Those include, in diesel engines, glowplugs for faster start-up; precombustion chambers (swirl chambers) for lower emissions, faster start-up and lower noise; turbocharger for reduced engine lag and emissions. In spark-ignition engines, silicon nitride is used for rocker arm pads for lower wear, turbocharger for lower inertia and less engine lag, and in exhaust gas control valves for increased acceleration. As examples of production levels, there is an estimated more than 300,000 sintered silicon nitride turbochargers made annually.

Bearings

Si_3N_4 bearing parts.

Silicon nitride bearings are both full ceramic bearings and ceramic hybrid bearings with balls in ceramics and races in steel. Silicon nitride ceramics have good shock resistance compared to other ceramics. Therefore, ball bearings made of silicon nitride ceramic are used in performance bearings. A representative example is use of silicon nitride bearings in the main engines of the NASA's Space Shuttle.

Since silicon nitride ball bearings are harder than metal, this reduces contact with the bearing track. This results in 80% less friction, 3 to 10 times longer lifetime, 80% higher speed, 60% less weight, the ability to operate with lubrication starvation, higher corrosion resistance and higher operation temperature, as compared to traditional metal bearings. Silicon nitride balls weigh 79% less than tungsten carbide balls. Silicon nitride ball bearings can be found in high end automotive bearings, industrial bearings, wind turbines, motorsports, bicycles, rollerblades and skateboards. Silicon nitride bearings are especially useful in applications where corrosion, electric or magnetic fields prohibit the use of metals. For example, in tidal flow meters, where seawater attack is a problem, or in electric field seekers.

Si_3N_4 was first demonstrated as a superior bearing in 1972 but did not reach production until nearly 1990 because of challenges associated with reducing the cost. Since 1990, the cost has been reduced substantially as production volume has increased. Although Si_3N_4 bearings are still 2–5 times more expensive than the best steel bearings, their superior performance and life are justifying rapid adoption. Around 15–20 million Si_3N_4 bearing balls were produced in the U.S. in 1996 for machine tools and many other applications. Growth is estimated at 40% per year, but could be even higher if ceramic bearings are selected for consumer applications such as in-line skates and computer disk drives.

High-temperature Material

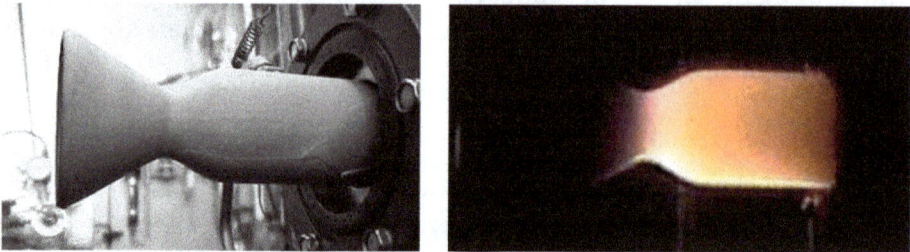

Silicon nitride thruster. Left: Mounted in test stand.
Right: Being tested with H_2/O_2 propellants.

Silicon nitride has long been used in high-temperature applications. In particular, it was identified as one of the few monolithic ceramic materials capable of surviving the severe thermal shock and thermal gradients generated in hydrogen/oxygen rocket engines. To demonstrate this capability in a complex configuration, NASA scientists used advanced rapid prototyping technology to fabricate a one-inch-diameter, single-piece combustion chamber/nozzle (thruster) component. The thruster was hot-fire tested with hydrogen/oxygen propellant and survived five cycles including a 5-minute cycle to a 1320 °C material temperature.

In 2010 silicon nitride was used as the main material in the thrusters of the JAXA space probe Akatsuki.

REFRACTORIES

Refractory is any material that has an unusually high melting point and that maintains its structural properties at very high temperatures. Composed principally of ceramics, refractories are employed in great quantities in the metallurgical, glass-making, and ceramics industries, where they are formed into a variety of shapes to line the interiors of furnaces, kilns, and other devices that process materials at high temperatures.

Properties

Because of the high strengths exhibited by their primary chemical bonds, many ceramics possess unusually good combinations of high melting point and chemical inertness. This makes them useful as refractories. (The word refractory comes from the French réfractaire, meaning "high-melting"). The property of chemical inertness is of special importance in metallurgy and glassmaking, where the furnaces are exposed to extremely corrosive molten materials and gases. In addition to temperature and corrosion resistance, refractories must possess superior physical wear or abrasion resistance, and they also must be resistant to thermal shock. Thermal shock occurs when an object is rapidly cooled from high temperature. The surface layers contract against the inner layers, leading to the development of tensile stress and the propagation of cracks. Ceramics, in spite of their well-known brittleness, can be made resistant to thermal shock by adjusting their microstructure during processing. The microstructure of ceramic refractories is quite coarse when compared with whitewares such as porcelain or even with less finely textured structural clay products such as brick. The size of filler grains can be on the scale of millimetres, instead of the micrometre scale seen in whiteware ceramics. In addition, most ceramic refractory products are quite porous, with large amounts of air spaces of varying size incorporated into the material. The presence of large grains and pores can reduce the load-bearing strength of the product, but it also can blunt cracks and thereby reduce susceptibility to thermal shock. However, in cases where a refractory will come into contact with corrosive substances (for example, in glass-melting furnaces), a porous structure is undesirable. The ceramic material can then be made with a higher density, incorporating smaller amounts of pores.

Composition and Processing

The composition and processing of ceramic refractories vary widely according to the application and the type of refractory. Most refractories can be classified on the basis of composition as either clay-based or nonclay-based. In addition, they can be classified as either acidic (containing silica [SiO_2] or zirconia [ZrO_2]) or basic (containing alumina [Al_2O_3] or alkaline-earth oxides such as lime [CaO] or magnesia [MgO]). Among the clay-based refractories are fireclay, high-alumina, and mullite ceramics. There is a wide range of nonclay refractories, including basic, extra-high alumina, silica, silicon carbide, and zircon materials. Most clay-based products are processed in a manner similar to other traditional ceramics such as structural clay products; e.g., stiff-mud processes such as press forming or extrusion are employed to form the ware, which is subsequently dried and passed through long tunnel kilns for firing. Firing, induces partial vitrification, or glass formation, which is a liquid-sinteringprocess that binds particles together. Nonclay-based refractories, on the other hand, are bonded using techniques reserved for advanced ceramic materials. For instance, extra-high alumina and zircon ceramics are bonded by transient-liquid or solid-state sintering, basic bricks are bonded by chemical reactions between constituents, and silicon carbide is reaction-bonded from silica sand and coke.

Clay-based Refractories

In this topic the composition and properties of the clay-based refractories are described. Most are produced as preformed brick. Much of the remaining products are so-called monolithics, materials that can be formed and solidified on-site. This category includes mortars for cementing bricks and mixes for ramming or gunning (spraying from a pressure gun) into place. In addition, lightweight refractory insulation can be made in the form of fibreboards, blankets, and vacuum-cast shapes.

Fireclay

The workhorse of the clay-based refractories are the so-called fireclay materials. These are made from clays containing the aluminosilicate mineral kaolinite($Al_2[Si_2O_5][OH]_4$) plus impurities such as alkalis and iron oxides. The alumina content ranges from 25 to 45 percent. Depending upon the impurity content and the alumina-to-silica ratio, fireclays are classified as low-duty, medium-duty, high-duty, and super-duty, with use temperature rising as alumina content increases. Fireclay bricks, or firebricks, exhibit relatively low expansion upon heating and are therefore moderately resistant against thermal shock. They are fairly inert in acidic environments but are quite reactive in basic environments. Fireclay bricks are used to line portions of the interiors of blast furnaces, blast-furnace stoves, and coke ovens.

High Alumina

High-alumina refractories are made from bauxite, a naturally occurring material containing aluminum hydroxide ($Al[OH]_3$) and kaolinitic clays. These raw materials are roasted to produce a mixture of synthetic alumina and mullite (an aluminosilicate mineral with the chemical formula $3Al_2O_3 \cdot 2SiO_2$). By definition high-alumina refractories contain between 50 and 87.5 percent alumina. They are much more robust than fireclay refractories at high temperatures and in basic environments. In addition, they exhibit better volume stability and abrasion resistance. High-alumina bricks are used in blast furnaces, blast-furnace stoves, and liquid-steel ladles.

Mullite

Mullite is an aluminosilicate compound with the specific formula $3Al_2O_3 \cdot 2SiO_3$ and an alumina content of approximately 70 percent. It has a melting point of 1,850 °C (3,360 °F). Various clays are mixed with bauxite in order to achieve this composition. Mullite refractories are solidified by sintering in electric furnaces at high temperatures. They are the most stable of the aluminosilicate refractories and have excellent resistance to high-temperature loading. Mullite bricks are used in blast-furnace stoves and in the forehearth roofs of glass-melting furnaces.

Non-clay-based Refractories

Nonclay refractories such as those described below are produced almost exclusively as bricks and pressed shapes, though some magnesite-chrome and alumina materials are fuse-cast into molds. The usual starting materials for these products are carbonates or oxides of metals such as magnesium, aluminum, and zirconium.

Basic refractories include magnesia, dolomite, chrome, and combinations of these materials. Magnesia brick is made from periclase, the mineral form of magnesia (MgO). Periclase is produced from magnesite (a magnesium carbonate, $MgCO_3$), or it is produced from magnesium hydroxide ($Mg[OH]_2$), which in turn is derived from seawater or underground brine solutions. Magnesia bricks can be chemically bonded, pitch-bonded, burned, or burned and then pitch-impregnated.

Dolomite refractories take their name from the dolomite ore, a combination of calcium and magnesium carbonates ($CaCO_3 \cdot MgCO_3$), from which they are produced. After burning they must be impregnated with tar or pitch to prevent rehydration of lime (CaO). Chrome brick is made from chromium ores, which are complex solid solutions of the spinel type (a series of oxide minerals including chromite and magnetite) plus silicate gangue, or impurity phases.

All the basic refractories exhibit outstanding resistance to iron oxides and the basic slags associated with steelmaking—especially when they incorporate carbon additions either as flakes or as residual carbon from pitch-bonding or tar-impregnation. For this reason they find wide employment in the linings of basic oxygen furnaces, electric furnaces, and open-hearth furnaces. They also are used to line the insides of copper converters.

Extra-high Alumina

Extra-high alumina refractories are classified as having between 87.5 and 100 percent Al_2O_3 content. The alumina grains are fused or densely sintered together to obtain high density. Extra-high alumina refractories exhibit excellent volume stability to over 1,800 °C (3,275 °F).

Silica

Silica refractories are made from quartzites and silica gravel deposits with low alumina and alkali contents. They are chemically bonded with 3–3.5 percent lime. Silica refractories have good load resistance at high temperatures, are abrasion-resistant, and are particularly suited to containing acidic slags. Of the various grades—coke-oven quality, conventional, and super-duty—the super-duty, which has particularly low impurity contents, is used in the superstructures of glass-melting furnaces.

Zircon

Refractories made of zircon (a zirconium silicate, $ZrSiO_4$) also are used in glasstanks because of their good resistance to the corrosive action of molten glasses. They possess good volume stability for extended periods at elevated temperatures, and they also show good creep resistance (i.e., low deformation under hot loading).

Silicon Carbide

Silicon carbide (SiC) ceramics are made by a process referred to as reaction bonding, invented by the American Edward G. Acheson in 1891. In the Acheson process, pure silica sand and finely divided carbon (coke) are reacted in an electric furnace at temperatures in the range of 2,200°–2,480 °C (4,000°–4,500 °F). SiC ceramics have outstanding high-temperature load-bearing strength and dimensional stability. They also exhibit great thermal shock resistance because of their high thermal conductivity. (In this case, high thermal conductivity prevents the formation of extreme temperature differences between inner and outer layers of a material, which frequently are a source of thermal expansion stresses.) Therefore, SiC makes good kiln furniture for supporting other ceramics during their firing.

Other non-clay-based Refractories

Other refractories produced in smaller quantities for special applications include graphite (a layered, multicrystalline form of carbon), zirconia (ZrO_2), forsterite(Mg_2SiO_4), and combinations such as magnesia-alumina, magnesite-chrome, chrome-alumina, and alumina-zirconia-silica. Alumina-zirconia-silica (AZS), which is melted and cast into molds or directly into the melting tanks of glass furnaces, is an excellent corrosion-resistant refractory that does not release impurities into the glass melt. AZS is also poured to make tank blocks (also called soldier blocks or sidewall blocks) used in the construction and repair of glass furnaces.

PORCELAIN

Porcelain is a ceramic material made by heating materials, generally including kaolin, in a kiln to temperatures between 1,200 and 1,400 °C (2,200 and 2,600 °F). The toughness, strength, and translucence of porcelain, relative to other types of pottery, arises mainly from vitrification and the formation of the mineral mullite within the body at these high temperatures. Though definitions vary, porcelain can be divided into three main categories: hard-paste, soft-paste and bone china. The category that an object belongs to depends on the composition of the paste used to make the body of the porcelain object and the firing conditions.

Soft-paste porcelain swan tureen, 1752–1756, Chelsea porcelain.

Porcelain slowly evolved in China and was finally achieved (depending on the definition used) at some point about 2,000 to 1,200 years ago, then slowly spread to other East Asian countries, and finally Europe and the rest of the world. Its manufacturing process is more demanding than that for earthenware and stoneware, the two other main types of pottery, and it has usually been regarded as the most prestigious type of pottery for its delicacy, strength, and its white colour. It combines well with both glazes and paint, and can be modelled very well, allowing a huge range of decorative treatments in tablewares, vessels and figurines. It also has many uses in technology and industry.

The European name, porcelain in English, comes from the old Italian *porcellana* (cowrie shell) because of its resemblance to the surface of the shell. Porcelain is also referred to as china or fine china in some English-speaking countries, as it was first seen in imports from China. Properties associated with porcelain include low permeability and elasticity; considerable strength, hardness, toughness, whiteness, translucency and resonance; and a high resistance to chemical attack and thermal shock.

Flower centrepiece, 18th century, Spain.

Porcelain has been described as being "completely vitrified, hard, impermeable (even before glazing), white or artificially coloured, translucent (except when of considerable thickness), and resonant". However, the term "porcelain" lacks a universal definition and has "been applied in an unsystematic fashion to substances of diverse kinds which have only certain surface-qualities in common".

Traditionally, East Asia only classifies pottery into low-fired wares (earthenware) and high-fired wares (often translated as porcelain), the latter also including what Europeans call stoneware, which is high-fired but not generally white or translucent. Terms such as "proto-porcelain", "porcellaneous" or "near-porcelain" may be used in cases where the ceramic body approaches whiteness and translucency.

Types

Hard Paste

Chinese Imperial Dish with Flowering Prunus, Famille
Rose overglaze enamel, between 1723 and 1735.

Hard-paste porcelain came from East Asia, specifically China, and some of the finest quality porcelain wares are from this category. The earliest European porcelains were produced at the Meissen factory in the early 18th century; they were formed from a paste composed of kaolin and alabaster and fired at temperatures up to 1,400 °C (2,552 °F) in a wood-fired kiln, producing a porcelain of great hardness, translucency, and strength. Later, the composition of the Meissen hard paste was changed and the alabaster was replaced by feldspar and quartz, allowing the pieces to be fired at lower temperatures. Kaolinite, feldspar and quartz (or other forms of silica) continue to constitute the basic ingredients for most continental European hard-paste porcelains.

Demonstration of the translucent quality of porcelain.

Soft Paste

Soft-paste porcelains date back from the early attempts by European potters to replicate Chinese porcelain by using mixtures of clay and frit. Soapstone and lime were known to have been included in these compositions. These wares were not yet actual porcelain wares as they were not hard nor vitrified by firing kaolin clay at high temperatures. As these early formulations suffered from high pyroplastic deformation, or slumping in the kiln at high temperatures, they were uneconomic to produce and of low quality.

Formulations were later developed based on kaolin with quartz, feldspars, nepheline syenite or other feldspathic rocks. These were technically superior, and continue to be produced. Soft-paste porcelains are fired at lower temperatures than hard-paste porcelain, therefore these wares are generally less hard than hard-paste porcelains.

Bone China

Although originally developed in England in 1748 in order to compete with imported porcelain, bone china is now made worldwide. The English had read the letters of Jesuit missionary François Xavier d'Entrecolles, which described Chinese porcelain manufacturing secrets in detail. One writer has speculated that a misunderstanding of the text could possibly have been responsible for the first attempts to use bone-ash as an ingredient of English porcelain, although this is not supported by researchers and historians.

Traditionally, English bone china was made from two parts of bone ash, one part of kaolin and one part china stone, although this has largely been replaced by feldspars from non-UK sources.

Materials

Kaolin is the primary material from which porcelain is made, even though clay minerals might account for only a small proportion of the whole. The word *paste* is an old term for both the unfired and fired materials. A more common terminology for the unfired material is "body"; for example, when buying materials a potter might order an amount of porcelain body from a vendor.

The composition of porcelain is highly variable, but the clay mineral kaolinite is often a raw material. Other raw materials can include feldspar, ball clay, glass, bone ash, steatite, quartz, petuntse and alabaster.

The clays used are often described as being long or short, depending on their plasticity. Long clays are cohesive (sticky) and have high plasticity; short clays are less cohesive and have lower plasticity. In soil mechanics, plasticity is determined by measuring the increase in content of water required to change a clay from a solid state bordering on the plastic, to a plastic state bordering on the liquid, though the term is also used less formally to describe the ease with which a clay may be worked.

Clays used for porcelain are generally of lower plasticity and are shorter than many other pottery clays. They wet very quickly, meaning that small changes in the content of water can produce large changes in workability. Thus, the range of water content within which these clays can be worked is very narrow and consequently must be carefully controlled.

Production

Forming

Porcelain can be made using all the shaping techniques for pottery. It was originally typically made on the potter's wheel, though moulds were also used from early on. Slip-casting has been the most common commercial method in recent times.

Glazing

Biscuit porcelain is unglazed porcelain treated as a finished product, mostly for figures and sculpture. Unlike their lower-fired counterparts, porcelain wares do not need glazing to render them impermeable to liquids and for the most part are glazed for decorative purposes and to make them resistant to dirt and staining. Many types of glaze, such as the iron-containing glaze used on the celadon wares of Longquan, were designed specifically for their striking effects on porcelain.

Decoration

Porcelain often receives underglaze decoration using pigments that include cobalt oxide and copper, or overglaze enamels, allowing a wider range of colours. Like many earlier wares, modern porcelains are often biscuit-fired at around 1,000 °C (1,830 °F), coated with glaze and then sent for a second glaze-firing at a temperature of about 1,300 °C (2,370 °F) or greater. Another early method is "once-fired", where the glaze is applied to the unfired body and the two fired together in a single operation.

Song dynasty celadon porcelain with a
fenghuang spout, 10th century, China.

Firing

In this process, "green" (unfired) ceramic wares are heated to high temperatures in a kiln to permanently set their shapes, vitrify the body and the glaze. Porcelain is fired at a higher temperature than earthenware so that the body can vitrify and become non-porous. Many types of porcelain in the past have been fired twice or even three times, to allow decoration using less robust pigments in overglaze enamel.

Other Uses

Electric Insulating Material

Porcelain insulator for medium-high voltage.

Porcelain and other ceramic materials have many applications in engineering, especially ceramic engineering. Porcelain is an excellent insulator for use at high voltage, especially in outdoor applications. Examples include: Terminals for high-voltage cables, bushings of power transformers, insulation of high frequency antennas and many other components.

Building Material

Dakin Building, Brisbane, California using porcelain panels.

Porcelain can be used as a building material, usually in the form of tiles or large rectangular panels. Modern porcelain tiles are generally produced by a number of recognised international standards and definitions. Manufacturers are found across the world with Italy being the global leader, producing over 380 million square metres in 2006. Historic examples of rooms decorated entirely in porcelain tiles can be found in several European palaces including ones at Galleria Sabauda in Turin, Museo di Doccia in Sesto Fiorentino, Museo di Capodimonte in Naples, the Royal Palace of Madrid and the nearby Royal Palace of Aranjuez. and the Porcelain Tower of Nanjing.

More recent noteworthy examples include the Dakin Building in Brisbane, California, and the Gulf Building in Houston, Texas, which when constructed in 1929 had a 21-metre-long (69 ft) porcelain logo on its exterior.

Bathroom Fittings

Porcelain Chamber Pots from Vienna.

Because of its durability, inability to rust and impermeability, glazed porcelain has been in use for personal hygiene since at least the third quarter of the 17th century. During this period, porcelain chamber pots were commonly found in higher-class European households, and the term "bourdaloue" was used as the name for the pot.

However bath tubs are not made of porcelain, but of porcelain enamel on a metal base, usually of cast iron. Porcelain enamel is a marketing term used in the US, and is not porcelain but vitreous enamel.

CERAMIC-IMPREGNATED FABRIC

Ceramic-impregnated fabric is a fabric that has been impregnated with ceramic. Nanometric bioceramic can be incorporated into the polymer from which the fabric is manufactured. Bioceramic nanoparticles are added to the fused polymer. Some types of ceramics show thermally-induced photoluminescence, emitting light in the far infrared (FIR) region of the electromagnetic spectrum. When in contact with the body heat, the thermoluminescence of the fabrics with embedded bioceramic is enhanced. Bioceramics presents high reflection coefficient for the infrared radiation.

Method of Production

One way in which fabrics can be impregnated with ceramic is the process of electrophoretic deposition or EPD in the industry. In this process, nanoceramic particles are put into a solution in which the fabric to be infiltrated will be placed. The solution is then heated to high temperatures and the fabric is placed into the solution. Next, a current is passed through and the nano ceramic particles coat and impregnate the fabric. The pH level as well as the amount of time and amount of current can affect how well the fabric is infiltrated and how it is coated. Other processing methods can be further broken down into what particles will be added. There are two distinct groups:

- The SiC group (which contains silicon, carbon as well as additives for oxidation processes). When using this group for the purpose of ceramic-impregnation, the textile must first undergo a treatment. This is usually pyrolytic carbon or BN, which is deposited using a chemical vapor infiltration (CVI). Next, an overlayer of SiC is deposited on the textile using the same method. After this step the matrix of the textile is infiltrated by a slurry made up of SiC particles which can either be put in a polymer or simply molten SiC. This process coats the entirety of the textile.

- The oxide group. Oxides commonly used are alumina, silica, mullite, and rare-earth phosphates. The process for impregnation is quite simple: a slurry is prepared with the oxide desired and the textile is placed in it. Again, the slurry can be a molten state or a polymer-based one.

There are a few differences between these groups. The SiC group has over twice the fracture strength and thermal conductivity when compared to the oxides group. However, the oxides are more stable in combustion and oxidizing environments. The process by which the textile is impregnated depends on what materials will be used, as well as the intended purpose of that fabric.

Uses

These fabrics are utilized for thermal, mechanical and electrical applications for a variety of reasons. Ceramic-impregnated fabrics are most importantly used in three main fields: aerospace, electronic, and industrial. In aerospace, the fabrics are used in space shuttles for the exit cone, door seals, micrometeorite shield, gaskets, booster access doors, shuttle tiles, and in the Whipple shield. Ceramic-impregnated fabrics are utilized in aerospace because they have low thermal conductivity and can be fabricated into high temperature thermal insulators. In the electronic industry, the fabrics are used primarily for insulation and seals, because of its low porosity. Ceramic fabric's industrial uses include furnace linings, furnace zone dividers, door seals, tube seals, gaskets, and expansion joints. In addition to being an effective thermal insulator, these fabrics do not shrink or elongate with high temperature changes, making them useful for industrial uses that involve high temperatures.

CERAMIC CEMENTS

Ceramic cements help isolate, insulate and protect components in a variety of applications.

Ceramic cements help to keep the different components that form a particular part isolated, insulated and protected.

Important physical characteristics of ceramic cements include the capacity to deal with temperature changes varying from ambient temperature up to 3000 °F (1650 °C) and higher.

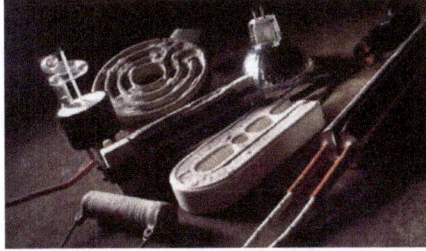

Specialty ceramic cements are directly relevant to our lives, but we rarely notice their presence. The important physical characteristics of these cements include the capacity to deal with temperature changes varying from ambient temperature up to 3000 °F (1650 °C) and higher, depending on the specific case.

Each material has a different coefficient of thermal expansion. Selecting the cement with the proper coefficient of thermal expansion (i.e., one that is similar to that of the other material in the application) is important to avoid cracks and failures.

Properties

Some ceramic cements are insulators; they are selected not only for their temperature resistance, but also for their ability to insulate and protect parts from damage due to extreme high temperatures, like in thermocouples used to check the quality of molten iron and other molten ores. Some ceramic cements are also chemical resistant, making them appropriate for service in acidic or alkaline environments. In other words, they are not only able to withstand the temperature changes and aggressive environments, but they can also be used as insulators or in contact with dissimilar materials. In addition, ceramic cements are able to isolate parts of an application from the surroundings. The combination of all of these properties helps the pieces and parts to do their jobs.

Nearly all ceramic cements are made of inorganic materials, which presents a small problem: porosity. Some ceramic cements are never going to be able to protect the part in full from water and humidity penetration. This is characteristic of inorganic ceramic cements. In many cases, the addition of a primer helps to delay or prevent this kind of penetration.

Some of these cements are made by mixing the cement with water, which is required for the material to cure. In the curing process, it is crucial to eliminate all the excess water (also known as water of convenience) used to hydrate the cement and initiate the curing process. After the cement has cured and reached its peak, that water of convenience needs to be removed before the part is put in service to avoid cracking, steaming or even short-circuiting the part (if it is transmitting electricity).

Some ceramic cements need to be fast curing to comply with manufacturers' demands to be able to assemble and pack the parts in order to send to consumers as soon as possible. Others need to be oven baked to cure properly, letting all the water used in the mixing of the cement reach a complete cure to avoid problems when the parts are put into service.

Selection

When selecting the proper cement, keep in mind that some ceramic cements might have a bad reaction when in contact with certain metals and other parts. To select the proper ceramic cement for your application, the ceramic supplier should be provided with the following details:

- What the application is.
- Degree of electrical resistance required.
- Temperature range the cement will withstand.
- Thermal shock resistance required.
- How fast the temperature goes from one extreme to the other.
- Frequency of those temperature changes.
- Desired thermal conductivity.
- Allowed degree of thermal expansion.
- What materials will be in contact with the cement.
- Whether volume stability is required.
- Whether moisture absorption will be a problem.
- Dispensing methods.
- Required pot life.
- Whether a predetermined set time exists.
- Whether a force cure is needed.

Ceramic cements can be applied in many different ways: potting, casting, sealing, encapsulating, bonding and coating. They help to keep the different components that form a particular part isolated, insulated and protected. They help to keep all the pieces together, bonding metals to ceramics and to glass, to avoid the de-bonding of the components.

Uses of Ceramic Cements

In Your Garage: Every breaker in your garage's electric panel is assembled using a specialty ceramic cement able to keep the calibration of the breaker in the right position. All of the international assemblers of those products use this kind of cement.

In Your Kitchen: In your kitchen, gas ranges with electric starters are all made with a special ceramic cement that can take the heat generated to ignite the gas for you to fry an egg or boil water. By pushing a button or turning a knob on your gas range, a spark jumps from the edge of the burner into the path from where the gas is coming to ignite the gas and provide flame for the burner.

The resistors located in the toaster are assembled using a specialty ceramic cement able to withstand the high temperature required to toast your bagel to perfection.

Many people believe these antique pieces are made entirely of silver, but the truth is that all of them are assembled using a special ceramic cement to give weight and balance to the piece of silverware and make it affordable to everyone.

In Your Car: Every light bulb on your dashboard is assembled with a specialty cement able to take the cycling of the temperature from the moment you start your car until you turn it off.

At the Mall: Those lamps are assembled using a ceramic cement that is able to handle not only the high temperature, but also the electrical conductivity necessary to keep the lamps lit.

TRANSPARENT CERAMICS

Transparent spinel ($MgAl_2O_4$) ceramic is used traditionally for applications such as high-energy laser windows because of its excellent transmission in visible wavelengths and mid-wavelength infrared (0.2-5.0μm) when combined with selected materials.

Many ceramic materials, both glassy and crystalline, have found use as optically transparent materials in various forms from bulk solid-state components to high surface area forms such as thin films, coatings, and fibers. Such devices have found widespread use for various applications in the electro-optical field including: Optical fibers for guided lightwave transmission, optical switches, laser amplifiers and lenses, hosts for solid-state lasers and optical window materials for gas lasers, and infrared (IR) heat seeking devices for missile guidance systems and IR night vision.

While single-crystalline ceramics may be largely defect-free (particularly within the spatial scale of the incident light wave), optical transparency in polycrystalline materials is limited by the amount of light that is scattered by their microstructural features. The amount of light scattering therefore depends on the wavelength of the incident radiation, or light.

For example, since visible light has a wavelength scale on the order of hundreds of nanometers, scattering centers will have dimensions on a similar spatial scale. Most ceramic materials, such as alumina and its compounds, are formed from fine powders, yielding a fine grained polycrystalline microstructure that is filled with scattering centers comparable to the wavelength of visible light. Thus, they are generally opaque as opposed to transparent materials. Recent nanoscale technology, however, has made possible the production of (poly)crystalline transparent ceramics such as alumina Al_2O_3, yttria alumina garnet (YAG), and neodymium-doped.

Synthetic sapphire - single-crystal aluminum oxide
(sapphire – Al_2O_3) is a transparent ceramic.

Transparent ceramics have recently acquired a high degree of interest and notoriety. Basic applications include lasers and cutting tools, transparent armor windows, night vision devices (NVD), and nose cones for heat seeking missiles. Currently available infrared (IR) transparent materials typically exhibit a trade-off between optical performance and mechanical strength. For example, sapphire (crystalline alumina) is very strong, but lacks full transparency throughout the 3–5 micrometer mid-IR range. Yttria is fully transparent from 3–5 micrometers, but lacks sufficient strength, hardness, and thermal shock resistance for high-performance aerospace applications. Not surprisingly, a combination of these two materials in the form of the yttria-alumina garnet (YAG) has proven to be one of the top performers in the field.

In 1961, General Electric began selling transparent alumina Lucalox bulbs. In 1966, GE announced a ceramic "transparent as glass," called Yttralox. In 2004, Anatoly Rosenflanz and colleagues at 3M used a "flame-spray" technique to alloy aluminium oxide (or alumina) with rare-earth metal oxides in order to produce high strength glass-ceramics with good optical properties. The method avoids many of the problems encountered in conventional glass forming and may be extensible to other oxides. This goal has been readily accomplished and amply demonstrated in laboratories and research facilities worldwide using the emerging chemical processing methods encompassed by the methods of sol-gel chemistry and nanotechnology.

Many ceramic materials, both glassy and crystalline, have found use as hosts for solid-state lasers and as optical window materials for gas lasers. The first working laser was made by Theodore H. Maiman in 1960 at Hughes Research Laboratories in Malibu, who had the edge on other research teams led by Charles H. Townes at Columbia University, Arthur Schawlow at Bell Labs, and Gould at TRG (Technical Research Group). Maiman used a solid-state light-pumped synthetic ruby to produce red laser light at a wavelength of 694 nanometers (nm). Synthethic ruby lasers are still in use. Both sapphires and rubies are corundum, a crystalline form of aluminium oxide (Al2O3).

Crystals

Ruby lasers consist of single-crystal sapphire alumina (Al_2O_3) rods doped with a small concentration of chromium Cr, typically in the range of 0.05%. The end faces are highly polished with a planar and parallel configuration. Neodymium-doped YAG (Nd:YAG) has proven to be one of the best solid-state laser materials. Its indisputable dominance in a broad variety of laser applications is determined by a combination of high emission cross section with long spontaneous emission lifetime, high damage threshold, mechanical strength, thermal conductivity, and low thermal beam distortion. The fact that the Czochralski crystal growth of Nd:YAG is a matured, highly reproducible and relatively simple technological procedure adds significantly to the value of the material.

Nd:YAG lasers are used in manufacturing for engraving, etching, or marking a variety of metals and plastics. They are extensively used in manufacturing for cutting and welding steel and various alloys. For automotive applications (cutting and welding steel) the power levels are typically 1–5 kW. In addition, Nd:YAG lasers are used in ophthalmology to correct posterior capsular opacification, a condition that may occur after cataract surgery, and for peripheral iridotomy in patients with acute angle-closure glaucoma, where it has superseded surgical iridectomy. Frequency-doubled Nd:YAG lasers (wavelength 532 nm) are used for pan-retinal photocoagulation in patients with diabetic retinopathy. In oncology, Nd:YAG lasers can be used to remove skin cancers. These lasers are also used extensively in the field of cosmetic medicine for laser hair removal and the treatment of minor vascular defects such as spider veins on the face and legs. Recently used for dissecting cellulitis, a rare skin disease usually occurring on the scalp. Using hysteroscopy in the field of gynecology, the Nd:YAG laser has been used for removal of uterine septa within the inside of the uterus. In dentistry, Nd:YAG lasers are used for soft tissue surgeries in the oral cavity.

Currently, high powered Nd:glass lasers as large as a football field
are used for inertial confinement fusion, nuclear weapons research,
and other high energy density physics experiments.

Glasses

Glasses (non-crystalline ceramics) also are used widely as host materials for lasers. Relative to crystalline lasers, they offer improved flexibility in size and shape and may be readily manufactured as large, homogeneous, isotropic solids with excellent optical properties. The indices of refraction of glass laser hosts may be varied between approximately 1.5 and 2.0, and both the temperature coefficient of n and the strain-optical coefficient may be tailored by altering the chemical composition. Glasses have lower thermal conductivities than the alumina or YAG, however, which imposes limitations on their use in continuous and high repetition-rate applications.

The principal differences between the behavior of glass and crystalline ceramic laser host materials are associated with the greater variation in the local environment of lasing ions in amorphous solids. This leads to a broadening of the fluorescent levels in glasses. For example, the width of the Nd^{3+} emission in YAG is ~ 10 angstroms as compared to ~ 300 angstroms in typical oxide glasses. The broadened fluorescent lines in glasses make it more difficult to obtain continuous wave laser operation (CW), relative to the same lasing ions in crystalline solid laser hosts.

Several glasses are used in transparent armor, such as normal plate glass (soda-lime-silica), borosilicate glass, and fused silica. Plate glass has been the most common glass used due to its low cost. But greater requirements for the optical properties and ballistic performance have necessitated the development of new materials. Chemical or thermal treatments can increase the strength of glasses, and the controlled crystallization of certain glass compositions can produce optical quality glass-ceramics. Alstom Grid Ltd. currently produces a lithium di-silicate based glass-ceramic known as TransArm, for use in transparent armor systems. It has all the workability of an amorphous glass, but upon recrystallization it demonstrates properties similar to a crystalline ceramic. Vycor is 96% fused silica glass, which is crystal clear, lightweight and high strength. One advantage of these type of materials is that they can be produced in large sheets and other curved shapes.

Nanomaterials

It has been shown fairly recently that laser elements (amplifiers, switches, ion hosts, etc.) made from fine-grained ceramic nanomaterials—produced by the low temperature sintering of high purity nanoparticles and powders—can be produced at a relatively low cost. These components are free of internal stress or intrinsic birefringence, and allow relatively large doping levels or optimized custom-designed doping profiles. This highlights the use of ceramic nanomaterials as being particularly important for high-energy laser elements and applications.

Primary scattering centers in polycrystalline nanomaterials—made from the sintering of high purity nanoparticles and powders—include microstructural defects such as residual porosity and grain boundaries. Thus, opacity partly results from the incoherent scattering of light at internal surfaces and interfaces. In addition to porosity, most of the interfaces or internal surfaces in ceramic nanomaterials are in the form of grain boundaries which separate nanoscale regions of crystalline order. Moreover, when the size of the scattering center (or grain boundary) is reduced well below the size of the wavelength of the light being scattered, the light scattering no longer occurs to any significant extent.

In the processing of high performance ceramic nanomaterials with superior opto-mechanical properties under adverse conditions, the size of the crystalline grains is determined largely by the size of the crystalline particles present in the raw material during the synthesis or formation of the object. Thus a reduction of the original particle size well below the wavelength of visible light (~ 0.5 μm or 500 nm) eliminates much of the light scattering, resulting in a translucent or even transparent material.

Furthermore, results indicate that microscopic pores in sintered ceramic nanomaterials, mainly trapped at the junctions of microcrystalline grains, cause light to scatter and prevented true transparency. It has been observed that the total volume fraction of these nanoscale pores (both intergranular and intragranular porosity) must be less than 1% for high-quality optical transmission, i.e. the density has to be 99.99% of the theoretical crystalline density.

Lasers

Nd:YAG

For example, a 1.46 kW Nd:YAG laser has been demonstrated by Konoshima Chemical Co. in Japan. In addition, Livermore researchers realized that these fine-grained ceramic nanomaterials might greatly benefit high-powered lasers used in the National Ignition Facility (NIF) Programs Directorate. In particular, a Livermore research team began to acquire advanced transparent nanomaterials from Konoshima to determine if they could meet the optical requirements needed for Livermore's Solid-State Heat Capacity Laser (SSHCL). Livermore researchers have also been testing applications of these materials for applications such as advanced drivers for laser-driven fusion power plants.

Assisted by several workers from the NIF, the Livermore team has produced 15 mm diameter samples of transparent Nd:YAG from nanoscale particles and powders, and determined the most important parameters affecting their quality. In these objects, the team largely followed the Japanese production and processing methodologies, and used an in house furnace to vacuum sinter the nanopowders. All specimens were then sent out for hot isostatic pressing (HIP). Finally, the components were returned to Livermore for coating and testing, with results indicating exceptional optical quality and properties.

One Japanese/East Indian consortium has focused specifically on the spectroscopic and stimulated emission characteristics of Nd^{3+} in transparent YAG nanomaterials for laser applications. Their materials were synthesized using vacuum sintering techniques. The spectroscopic studies suggest overall improvement in absorption and emission and reduction in scattering loss. Scanning electron microscope and transmission electron microscope observations revealed an excellent optical quality with low pore volume and narrow grain boundary width. Fluorescence and Raman measurements reveal that the Nd^{3+} doped YAG nanomaterial is comparable in quality to its single-crystal counterpart in both its radiative and non-radiative properties. Individual Stark levels are obtained from the absorption and fluorescence spectra and are analyzed in order to identify the stimulated emission channels possible in the material. Laser performance studies favor the use of high dopant concentration in the design of an efficient microchip laser. With 4 at% dopant, the group obtained a slope efficiency of 40%. High-power laser experiments yield an optical-to-optical conversion efficiency of 30% for Nd (0.6 at%) YAG nanomaterial as compared to 34% for an Nd (0.6 at%) YAG single crystal. Optical gain measurements conducted in these materials also show values comparable to single crystal, supporting the contention that these materials could be suitable substitutes to single crystals in solid-state laser applications.

Yttria, Y_2O_3

The initial work in developing transparent yttrium oxide nanomaterials was carried out by General Electric in the 1960s.

IR 100 Award, Yttralox.

In 1966, a transparent ceramic, Yttralox, was invented by Dr. Richard C. Anderson at the General Electric Research Laboratory, with further work at GE's Metallurgy and Ceramics Laboratory by Drs. Paul J. Jorgensen, Joseph H. Rosolowski, and Douglas St. Pierre. Yttralox is "transparent as glass," has a melting point twice as high, and transmits frequencies in the near infrared band as well as visible light.

Gemstones of Yttralox transparent ceramic.

Richard C. Anderson holding a sample of Yttralox.

Further development of yttrium ceramic nanomaterials was carried out by General Electric in the 1970s in Schenectady and Cleveland, motivated by lighting and ceramic laser applications. Yttralox, transparent yttrium oxide Y_2O_3 containing ~ 10% thorium oxide (ThO_2) was fabricated by Greskovich and Woods. The additive served to control grain growth during densification, so that porosity remained on grain boundaries and not trapped inside grains where it would be quite difficult to eliminate during the initial stages of sintering. Typically, as polycrystalline ceramics densify during heat treatment, grains grow in size while the remaining porosity decreases both in volume fraction and in size. Optically transparent ceramics must be virtually pore-free.

GE's transparent Yttralox was followed by GTE's lanthana-doped yttria with similar level of additive. Both of these materials required extended firing times at temperatures above 2000 °C. La_2O_3 – doped Y_2O_3 is of interest for infrared (IR) applications because it is one of the longest wavelength transmitting oxides. It is refractory with a melting point of 2430 °C and has a moderate coefficient of thermal expansion coefficient. The thermal shock and erosion resistance is considered to be intermediate among the oxides, but outstanding compared to non-oxide IR transmitting materials. A major consideration is the low emissivity of yttria, which limits background radiation upon heating. It is also known that the phonon edge gradually moves to shorter wavelengths as a material is heated.

In addition, ytrria itself, Y_2O_3 has been clearly identified as a prospective solid-state laser material. In particular, lasers with ytterbium as dopant allow the efficient operation both in cw operation and in pulsed regimes.

At high concentration of excitations (of order of 1%) and poor cooling, the quenching of emission at laser frequency and avalanche broadband emission takes place.

The Livermore team is also exploring new ways to chemically synthesize the initial nanopowders. Borrowing on expertise developed in CMS over the past 5 years, the team is synthesizing nanopowders based on sol-gel processing, and then sintering them accordingly in order to obtain the solid-state laser components. Another technique being tested utilizes a combustion process in order to generate the powders by burning an organic solid containing yttrium, aluminum, and neodymium. The smoke is then collected, which consists of spherical nanoparticles.

The Livermore team is also exploring new forming techniques (e.g. extrusion molding) which have the capacity to create more diverse, and possibly more complicated, shapes. These include shells and tubes for improved coupling to the pump light and for more efficient heat transfer. In addition, different materials can be co-extruded and then sintered into a monolithic transparent solid. An amplifier slab can formed so that part of the structure acts in guided lightwave transmission in order to focus pump light from laser diodes into regions with a high concentration of dopant ions near the slab center.

In general, nanomaterials promise to greatly expand the availability of low-cost, high-end laser components in much larger sizes than would be possible with traditional single crystalline ceramics. Many classes of laser designs could benefit from nanomaterial-based laser structures such as amplifies with built-in edge claddings. Nanomaterials could also provide more robust and compact designs for high-peak power, fusion-class lasers for stockpile stewardship, as well as high-average-power lasers for global theater ICBM missile defense systems (e.g. Strategic Defense Initiative SDI, or more recently the Missile Defense Agency.

Night Vision

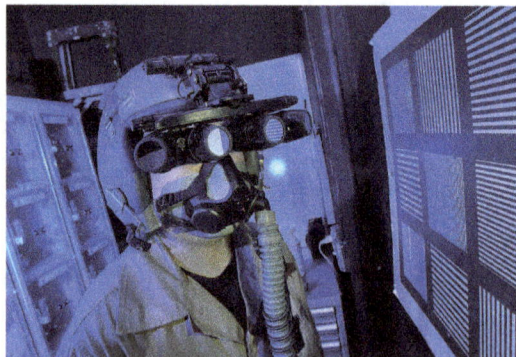

Panoramic Night Vision Goggles in testing.

A night vision device (NVD) is an optical instrument that allows images to be produced in levels of light approaching total darkness. They are most often used by the military and law enforcement agencies, but are available to civilian users. Night vision devices were first used in World War II, and came into wide use during the Vietnam War. The technology has evolved greatly since their introduction, leading to several "generations" of night vision equipment with performance increasing and price decreasing. The United States Air Force is experimenting with Panoramic Night Vision Goggles (PNVGs) which double the user's field of view to approximately 95 degrees by using four 16 mm image intensifiers tubes, rather than the more standard two 18 mm tubes.

Thermal images are visual displays of the amount of infrared (IR) energy emitted, transmitted, and reflected by an object. Because there are multiple sources of the infrared energy, it is difficult to get an accurate temperature of an object using this method. A thermal imaging camera is capable of performing algorithms to interpret that data and build an image. Although the image shows the viewer an approximation of the temperature at which the object is operating, the camera is using multiple sources of data based on the areas surrounding the object to determine that value rather than detecting the temperature.

Night vision infrared devices image in the near-infrared, just beyond the visual spectrum, and can see emitted or reflected near-infrared in complete visual darkness. All objects above the absolute zero temperature (0 K) emit infrared radiation. Hence, an excellent way to measure thermal variations is to use an infrared vision device, usually a focal plane array (FPA) infrared camera capable of detecting radiation in the mid (3 to 5 µm) and long (7 to 14 µm) wave infrared bands, denoted as MWIR and LWIR, corresponding to two of the high transmittance infrared windows. Abnormal temperature profiles at the surface of an object are an indication of a potential problem. Infrared thermography, thermal imaging, and thermal video, are examples of infrared imaging science. Thermal imaging cameras detect radiation in the infrared range of the electromagnetic spectrum (roughly 900–14,000 nanometers or 0.9–14 µm) and produce images of that radiation, called *thermograms.*

Since infrared radiation is emitted by all objects near room temperature, according to the black body radiation law, thermography makes it possible to see one's environment with or without visible illumination. The amount of radiation emitted by an object increases with temperature. Therefore, thermography allows one to see variations in temperature. When viewed through a thermal imaging camera, warm objects stand out well against cooler backgrounds; humans and other warm-blooded animals become easily visible against the environment, day or night. As a result, thermography is particularly useful to the military and to security services.

Thermography

In thermographic imaging, infrared radiation with wavelengths between 8–13 micrometers strikes the detector material, heating it, and thus changing its electrical resistance. This resistance change is measured and processed into temperatures which can be used to create an image. Unlike other types of infrared detecting equipment, microbolometers utilizing a transparent ceramic detector do not require cooling. Thus, a microbolometer is essentially an uncooled thermal sensor.

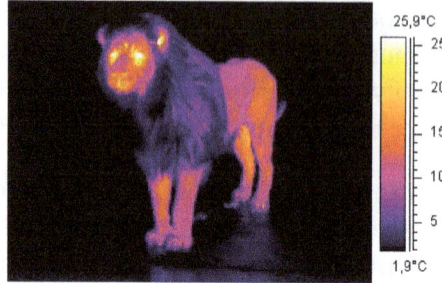

Thermogram of a lion.

The material used in the detector must demonstrate large changes in resistance as a result of minute changes in temperature. As the material is heated, due to the incoming infrared radiation, the resistance of the material decreases. This is related to the material's temperature coefficient of resistance (TCR) specifically its negative temperature coefficient. Industry currently manufactures microbolometers that contain materials with TCRs near −2%.

VO_2 and V_2O_5

The most commonly used ceramic material in IR radiation microbolometers is vanadium oxide. The various crystalline forms of vanadium oxide include both VO_2 and V_2O_5. Deposition at high temperatures and performing post-annealing allows for the production of thin films of these crystlalline compounds with superior properties, which may be easily integrated into the fabrication process. VO_2 has low resistance but undergoes a metal-insulator phase change near 67 °C and also has a lower TCR value. On the other hand, V_2O_5 exhibits high resistance and also high TCR.

Other IR transparent ceramic materials that have been investigated include doped forms of CuO, MnO and SiO.

Missiles

Many ceramic nanomaterials of interest for transparent armor solutions are also used for electromagnetic (EM) windows. These applications include radomes, IR domes, sensor protection, and multi-spectral windows. Optical properties of the materials used for these applications are critical, as the transmission window and related cut-offs (UV − IR) control the spectral bandwidth over which the window is operational. Not only must these materials possess abrasion resistance and strength properties common of most armor applications, but due to the extreme temperatures associated with the environment of military aircraft and missiles, they must also possess excellent thermal stability.

Thermal radiation is electromagnetic radiation emitted from the surface of an object which is due to the object's temperature. Infrared homing refers to a passive missile guidance system which uses the emission from a target of electromagnetic radiation in the infrared part of the spectrum to track it. Missiles that use infrared seeking are often referred to as "heat-seekers", since infrared is just below the visible spectrum of light in frequency and is radiated strongly by hot bodies. Many objects such as people, vehicle engines and aircraft generate and retain heat, and as such, are especially visible in the infrared wavelengths of light compared to objects in the background.

Sapphire

The current material of choice for high-speed infrared-guided missile domes is single-crystal sapphire. The optical transmission of sapphire does not extend to cover the entire mid-infrared range (3–5 μm), but starts to drop off at wavelengths greater than approximately 4.5 μm at room temperature. While the strength of sapphire is better than that of other available mid-range infrared dome materials at room temperature, it weakens above ~600 °C.

Limitations to larger area sapphires are often business related, in that larger induction furnaces and costly tooling dies are necessary in order to exceed current fabrication limits. However, as an industry, sapphire producers have remained competitive in the face of coating-hardened glass and new ceramic nanomaterials, and still managed to offer high performance and an expanded market.

Alternative materials, such as yttrium oxide, offer better optical performance, but inferior mechanical durability. Future high-speed infrared-guided missiles will require new domes that are substantially more durable than those in use today, while still retaining maximum transparency across a wide wavelength range. A long-standing trade-off exists between optical bandpass and mechanical durability within the current collection of single-phase infrared transmitting materials, forcing missile designers to compromise on system performance. Optical nanocomposites may present the opportunity to engineer new materials that overcome this traditional compromise.

The first full scale missile domes of transparent yttria manufactured from nanoscale ceramic powders were developed in the 1980s under Navy funding. Raytheon perfected and characterized its undoped polycrystalline yttria, while lanthana-doped yttria was similarly developed by GTE Labs. The two versions had comparable IR transmittance, fracture toughness, and thermal expansion, while the undoped version exhibited twice the value of thermal conductivity.

Renewed interest in yttria windows and domes has prompted efforts to enhance mechanical properties by using nanoscale materials with submicrometer or nano-sized grains. In one study, three vendors were selected to provide nanoscale powders for testing and evaluation, and they were compared to a conventional (5 μm) yttria powder previously used to prepare transparent yttria. While all of the nano-powders evaluated had impurity levels that were too high to allow processing to full transparency, 2 of them were processed to theoretical density and moderate transparency. Samples were sintered to a closed pore state at temperatures as low as 1400 °C.

After the relatively short sintering period, the component is placed in a hot isostatic press (HIP) and processed for 3–10 hours at ~ 30 kpsi (~200 MPa) at a temperature similar to that of the initial sintering. The applied isostatic pressure provides additional driving force for densification by substantially increasing the atomic diffusion coefficients, which promotes additional viscous flow at or near grain boundaries and intergranular pores. Using this method, transparent yttria nanomaterials were produced at lower temperatures, shorter total firing times, and without extra additives which tend to reduce the thermal conductivity.

Recently, a newer method has been developed by Mouzon, which relies on the methods of glass-encapsulation, combined with vacuum sintering at 1600 °C followed by hot isostatic pressing (HIP) at 1500 °C of a highly agglomerated commercial powder. The use of evacuated glass capsules to perform HIP treatment allowed samples that showed open porosity after vacuum sintering to be sintered to transparency. The sintering response of the investigated powder was studied by careful microstructural observations using scanning electron microscopy and optical microscopy both in reflection and transmission. The key to this method is to keep porosity intergranular during pre-sintering, so that it can be removed subsequently by HIP treatment. It was found that agglomerates of closely packed particles are helpful to reach that purpose, since they densify fully and leave only intergranular porosity.

Composites

Prior to the work done at Raytheon, optical properties in nanocomposite ceramic materials had received little attention. Their studies clearly demonstrated near theoretical transmission in nanocomposite optical ceramics for the first time. The yttria/magnesia binary system is an ideal model system for nanocomposite formation. There is limited solid solubility in either one of the constituent phases, permitting a wide range of compositions to be investigated and compared to each other. According to the phase diagram, bi-phase mixtures are stable for all temperatures below ~ 2100 °C. In addition, neither yttria nor magnesia shows any absorption in the $3 - 5$ μm mid-range IR portion of the EM spectrum.

In optical nanocomposites, two or more interpenetrating phases are mixed in a sub-micrometer grain sized, fully dense body. Infrared light scattering can be minimized (or even eliminated) in the material as long as the grain size of the individual phases is significantly smaller than infrared wavelengths. Experimental data suggests that limiting the grain size of the nanocomposite to approximately 1/15th of the wavelength of light is sufficient to limit scattering.

Nanocomposites of yttria and magnesia have been produced with a grain size of approximately 200 nm. These materials have yielded good transmission in the 3–5 μm range and strengths higher than that for single-phase individual constituents. Enhancement of mechanical properties in nanocomposite ceramic materials has been extensively studied. Significant increases in strength (2–5 times), toughness (1–4 times), and creep resistance have been observed in systems including SiC/Al_2O_3, SiC/Si_3N_4, SiC/MgO, and Al_2O_3/ZrO_2.

The strengthening mechanisms observed vary depending on the material system, and there does not appear to be any general consensus regarding strengthening mechanisms, even within a given system. In the SiC/Al_2O_3 system, for example, it is widely known and accepted that the addition of SiC particles to the Al_2O_3 matrix results in a change of failure mechanism from intergranular (between grains) to intragranular (within grains) fracture. The explanations for improved strength include:

- A simple reduction of processing flaw concentration during nanocomposite fabrication.

- Reduction of the critical flaw size in the material—resulting in increased strength as predicted by the Hall-Petch relation).

- Crack deflection at nanophase particels due to residual thermal stresses introduced upon cooling form processing temperatures.

- Microcracking along stress-induced dislocations in the matrix material.

Armor

There is an increasing need in the military sector for high-strength, robust materials which have the capability to transmit light around the visible (0.4–0.7 micrometers) and mid-infrared (1–5 micrometers) regions of the spectrum. These materials are needed for applications requiring transparent armor. Transparent armor is a material or system of materials designed to be optically transparent, yet protect from fragmentation or ballistic impacts. The primary requirement for a transparent armor system is to not only defeat the designated threat but also provide a multi-hit capability with minimized distortion of surrounding areas. Transparent armor windows must also be compatible with night vision equipment. New materials that are thinner, lightweight, and offer better ballistic performance are being sought.

Existing transparent armor systems typically have many layers, separated by polymer (e.g. polycarbonate) interlayers. The polymer interlayer is used to mitigate the stresses from thermal expansion mismatches, as well as to stop crack propagation from ceramic to polymer. The polycarbonate is also currently used in applications such as visors, face shields and laser protection goggles. The search for lighter materials has also led to investigations into other polymeric materials such as transparent nylons, polyurethane, and acrylics. The optical properties and durability of transparent plastics limit their use in armor applications. Investigations carried out in the 1970s had shown promise for the use of polyurethane as armor material, but the optical properties were not adequate for transparent armor applications.

Several glasses are utilized in transparent armor, such as normal plate glass (soda-lime-silica), borosilicate glasses, and fused silica. Plate glass has been the most common glass used due to its low cost, but greater requirements for the optical properties and ballistic performance have generated the need for new materials. Chemical or thermal treatments can increase the strength of glasses, and the controlled crystallization of certain glass systems can produce transparent glass-ceramics. Alstom Grid Research & Technology, produced a lithium disilicate based glass-ceramic known as TransArm, for use in transparent armor systems with continuous production yielding vehicle windscreen sized pieces (and larger). The inherent advantages of glasses and glass-ceramics include having lower cost than most other ceramic materials, the ability to be produced in curved shapes, and the ability to be formed into large sheets.

Transparent crystalline ceramics are used to defeat advanced threats. Three major transparent candidates currently exist: aluminum oxynitride (AlON), magnesium aluminate spinel (spinel), and single crystal aluminum oxide (sapphire).

Aluminium Oxynitride Spinel

Aluminium oxynitride spinel ($Al_{23}O_{27}N_5$), abbreviated as AlON, is one of the leading candidates for transparent armor. It is produced by the Surmet Corporation under the trademark ALON. The incorporation of nitrogen into aluminium oxide stabilizes a crystalline spinel phase, which due to its cubic crystal structure and unit cell, is an isotropic material which can be produced as transparent ceramic nanomaterial. Thus, fine-grained polycrystalline nanomaterials can be produced and formed into complex geometries using conventional ceramic forming techniques such as hot isostatic pressing, and slip casting.

The Surmet Corporation has acquired Raytheon's ALON business and is currently building a market for this technology in the area of Transparent Armor, Sensor windows, Reconnaissance windows and IR Optics such as Lenses and Domes and as an alternative to quartz and sapphire in the semiconductor market. The AlON based transparent armor has been tested to stop multi-hit threats including of 30calAPM2 rounds and 50calAPM2 rounds successfully. The high hardness of AlON provides a scratch resistance which exceeds even the most durable coatings for glass scanner windows, such as those used in supermarkets. Surmet has successfully produced a 15" x 18" curved AlON window and is currently attempting to scale up the technology and reduce the cost. In addition, the U.S. Army and U.S. Air Force are both seeking development into next generation applications.

Spinel

Magnesium aluminate spinel ($MgAl_2O_4$) is a transparent ceramic with a cubic crystal structure with an excellent optical transmission from 0.2 to 5.5 micrometers in its polycrystalline form. Optical quality transparent spinel has been produced by sinter/ HIP, hot pressing, and hot press/HIP operations, and it has been shown that the use of a hot isostatic press can improve its optical and physical properties.

Spinel offers some processing advantages over AlON, such as the fact that spinel powder is available from commercial manufacturers while AlON powders are proprietary to Raytheon. It is also capable of being processed at much lower temperatures than AlON and has been shown to possess superior optical properties within the infrared (IR) region. The improved optical characteristics make spinel attractive in sensor applications where effective communication is impacted by the protective missile dome's absorption characteristics.

Spinel shows promise for many applications, but is currently not available in bulk form from any manufacturer, although efforts to commercialize spinel are underway. The spinel products business is being pursued by two key U.S. manufacturers: "Technology Assessment and Transfer" and the "Surmet Corporation".

An extensive NRL review of the literature has indicated clearly that attempts to make high-quality spinel have failed to date because the densification dynamics of spinel are poorly understood. They have conducted extensive research into the dynamics involved during the densification of spinel. Their research has shown that LiF, although necessary, also has extremely adverse effects during the final stages of densification. Additionally, its distribution in the precursor spinel powders is of critical importance.

Traditional bulk mixing processes used to mix LiF sintering aid into a powder leave fairly inhomogeneous distribution of Lif that must be homogenized by extended heat treatment times at elevated temperatures. The homogenizing temperature for Lif/Spinel occurs at the temperature of fast reaction between the LiF and the Al_2O_3. In order to avoid this detrimental reaction, they have developed a new process that uniformly coats the spinel particles with the sintering aid. This allows them to reduce the amount of Lif necessary for densification and to rapidly heat through the temperature of maximum reactivity. These developments have allowed NRL to fabricate $MgAl_2O_4$ spinel to high transparency with extremely high reproducibility that should enable military as well as commercial use of spinel.

Sapphire

Single-crystal aluminum oxide (sapphire – Al_2O_3) is a transparent ceramic. Sapphire's crystal structure is rhombohedral and thus its properties are anisotropic, varying with crystallographic orientation. Transparent alumina is currently one of the most mature transparent ceramics from a production and application perspective, and is available from several manufacturers. But the cost is high due to the processing temperature involved, as well as machining costs to cut parts out of single crystal boules. It also has a very high mechanical strength – but that is dependent on the surface finish.

The high level of maturity of sapphire from a production and application standpoint can be attributed to two areas of business: Electromagnetic spectrum windows for missiles and domes, and electronic/semiconductor industries and applications.

There are current programs to scale-up sapphire grown by the heat exchanger method or edge defined film-fed growth (EFG) processes. Its maturity stems from its use as windows and in semiconductor industry. Crystal Systems Inc. which uses single crystal growth techniques, is currently scaling their sapphire boules to 13-inch (330 mm) diameter and larger. Another producer, the Saint-Gobain Group produces transparent sapphire using an edge-defined growth technique. Sapphire grown by this technique produces an optically inferior material to that which is grown via single crystal techniques, but is much less expensive, and retains much of the hardness, transmission, and scratch-resistant characteristics. Saint-Gobain is currently capable of producing 0.43' thick (as grown) sapphire, in 12' × 18.5' sheets, as well as thick, single-curved sheets. The U.S. Army Research Laboratory is currently investigating use of this material in a laminate design for transparent armor systems. The Saint Gobain Group have commercialized the capability to meet flight requirements on the F-35 Joint Strike Fighter and F-22 Raptor next generation fighter aircraft.

Composites

Future high-speed infrared-guided missiles will require new dome materials that are substantially more durable than those in use today, while retaining maximum transparency across the entire operational spectrum or bandwidth. A long-standing compromise exists between optical bandpass and mechanical durability within the current group of single-phase (crystalline or glassy) IR transmitting ceramic materials, forcing missile designers to accept substandard overall system performance. Optical nanocomposites may provide the opportunity to engineer new materials that may overcome these traditional limitations.

For example, transparent ceramic armor consisting of a lightweight composite has been formed by utilizing a face plate of transparent alumina Al_2O_3 (or magnesia MgO) with a back-up plate of transparent plastic. The two plates (bonded together with a transparent adhesive) afford complete ballistic protection against 0.30 AP M2 projectiles at 0° obliquity with a muzzle velocity of 2,770 ft (840 m) per second. Another transparent composite armor provided complete protection for small arms projectiles up to and including caliber .50 AP M2 projectiles consisting of two or more layers of transparent ceramic material.

Nanocomposites of yttria and magnesia have been produced with an average grain size of ~200 nm. These materials have exhibited near theoretical transmission in the 3 − 5 µm IR band. Additionally, such composites have yielded higher strengths than those observed for single phase solid-state components. Despite a lack of agreement regarding mechanism of failure, it is widely accepted that nanocomposite ceramic materials can and do offer improved mechanical properties over those of single phase materials or nanomaterials of uniform chemical composition.

Nanocomposite ceramic materials also offer interesting mechanical properties not achievable in other materials, such as superplastic flow and metal-like machinability. It is anticipated that further development will result in high strength, high transparency nanomaterials which are suitable for application as next generation armor.

TERRACOTTA

Terracotta, terra cotta or terra-cotta, a type of earthenware, is a clay-based unglazed or glazed ceramic, where the fired body is porous. Terracotta is the term normally used for sculpture made in earthenware, and also for various practical uses including vessels (notably flower pots), water and waste water pipes, roofing tiles, bricks, and surface embellishment in building construction. The term is also used to refer to the natural brownish orange color of most terracotta, which varies considerably.

Asian and European sculpture in porcelain is not covered. Glazed architectural terracotta and its unglazed version as exterior surfaces for buildings were used in Asia for some centuries before becoming popular in the West in the 19th century. Architectural terracotta can also refer to decorated ceramic elements such as antefixes and revetments, which made a large contribution to the appearance of temples and other buildings in the classical architecture of Europe, as well as in the Ancient Near East.

In archaeology and art history, "terracotta" is often used to describe objects such as figurines not made on a potter's wheel. Vessels and other objects that are or might be made on a wheel from the same material are called earthenware pottery; the choice of term depends on the type of object rather than the material or firing technique. Unglazed pieces, and those made for building construction and industry, are also more likely to be referred to as terracotta, whereas tableware and other vessels are called earthenware (though sometimes terracotta if unglazed), or by a more precise term such as faience.

In Art

Terracotta female figurines were uncovered by archaeologists in excavations of Mohenjo-daro, Pakistan. Along with phallus-shaped stones, these suggest some sort of fertility cult. The Burney Relief is an outstanding terracotta plaque from Ancient Mesopotamia of about 1950 BC. In Mesoamerica, the great majority of Olmec figurines were in terracotta. Many ushabti mortuary statuettes were also made of terracotta in Ancient Egypt.

The Ancient Greeks' Tanagra figurines were mass-produced mold-cast and fired terracotta figurines, that seem to have been widely affordable in the Hellenistic period, and often purely decorative in function. They were part of a wide range of Greek terracotta figurines, which included larger and higher-quality works such as the Aphrodite Heyl; the Romans too made great numbers of small figurines, often religious. Etruscan art often used terracotta in preference to stone even for larger statues, such as the near life-size Apollo of Veii and the *Sarcophagus of the Spouses*. Campana reliefs are Ancient Roman terracotta reliefs, originally mostly used to make friezes for the outside of buildings, as a cheaper substitute for stone.

Indian sculpture made heavy use of terracotta from as early as the Indus Valley Civilization (with stone and metal sculpture being rather rare), and in more sophisticated areas had largely abandoned modeling for using molds by the 1st century BC. This allows relatively large figures, nearly up to life-size, to be made, especially in the Gupta period and the centuries immediately following it. Several vigorous local popular traditions of terracotta folk sculpture remain active today, such as the Bankura horses.

Precolonial West African sculpture also made extensive use of terracotta. The regions most recognized for producing terracotta art in that part of the world include the Nok culture of central and north-central Nigeria, the Ife/Benin cultural axis in western and southern Nigeria (also noted for its exceptionally naturalistic sculpture), and the Igbo culture area of eastern Nigeria, which excelled in terracotta pottery. These related, but separate, traditions also gave birth to elaborate schools of bronze and brass sculpture in the area.

Chinese sculpture made great use of terracotta, with and without glazing and colour, from a very early date. The famous Terracotta Army of Emperor Qin Shi Huang, 209–210 BC, was somewhat untypical, and two thousand years ago reliefs were more common, in tombs and elsewhere. Later Buddhist figures were often made in painted and glazed terracotta, with the Yixian glazed pottery luohans, probably of 1150–1250, now in various Western museums, among the finest examples. Brick-built tombs from the Han dynasty were often finished on the interior wall with bricks decorated on one face; the techniques included molded reliefs. Later tombs contained many figures of protective spirits and animals and servants for the afterlife, including the famous horses of the T'ang dynasty; as an arbitrary matter of terminology these tend not to be referred to as terracottas.

European medieval art made little use of terracotta sculpture, until the late 14th century, when it became used in advanced International Gothic workshops in parts of Germany. A few decades later there was a revival in the Italian Renaissance, inspired by excavated classical terracottas as well as the German examples, which gradually spread to the rest of Europe. In Florence Luca della Robbia was a sculptor who founded a family dynasty specializing in glazed and painted terracotta, especially large roundels which were used to decorate the exterior of churches and other buildings. These used the same techniques as contemporary maiolica and other tin-glazed pottery. Other sculptors included Pietro Torrigiano, who produced statues, and in England busts of the Tudor royal family. The unglazed busts of the Roman Emperors adorning Hampton Court Palace, by Giovanni da Maiano, 1521, were another example of Italian work in England. They were originally painted but this has now been lost from weathering.

Clodion, *The River Rhine Separating the Waters.*

In the 18th-century unglazed terracotta, which had long been used for preliminary clay models or maquettes that were then fired, became fashionable as a material for small sculptures including portrait busts. It was much easier to work than carved materials, and allowed a more spontaneous approach by the artist. Claude Michel, known as Clodion, was an influential pioneer in France. John Michael Rysbrack, a Flemish portrait sculptor working in England, sold his terracotta *modelli* for larger works in stone, and produced busts only in terracotta. In the next century the French sculptor Albert-Ernest Carrier-Belleuse made many terracotta pieces, but possibly the most famous is The Abduction of Hippodameia depicting the Greek mythological scene of a centaur kidnapping Hippodameia on her wedding day.

Architecture

Imperial roof decoration in the Forbidden City.

Many ancient and traditional roofing styles included more elaborate sculptural elements than the plain roof tiles, such as Chinese Imperial roof decoration and the antefix of western classical architecture. In India West Bengal made a speciality of terracotta temples, with the sculpted decoration from the same material as the main brick construction.

In the 19th century the possibilities of terracotta decoration of buildings were again appreciated by architects, often using thicker pieces of terracotta, and surfaces that are not flat. The American architect Louis Sullivan is well known for his elaborate glazed terracotta ornamentation, designs that would have been impossible to execute in any other medium. Terracotta and tile were used extensively in the town buildings of Victorian Birmingham, England. By about 1930 the widespread use of concrete and Modernist architecture largely ended the use of terracotta in architecture.

Production and Properties

An appropriate refined clay is formed to the desired shape. After drying it is placed in a kiln or atop combustible material in a pit, and then fired. The typical firing temperature is around 1,000 °C (1,830 °F), though it may be as low as 600 °C (1,112 °F) in historic and archaeological examples. The iron content, reacting with oxygen during firing, gives the fired body a reddish color, though the overall color varies widely across shades of yellow, orange, buff, red, "terracotta", pink, grey or brown. In some contexts, such as Roman figurines, white-colored terracotta is known as pipeclay, as such clays were later preferred for tobacco pipes, normally made of clay until the 19th century.

Fired terracotta is not watertight, but surface-burnishing the body before firing can decrease its porousness and a layer of glaze can make it watertight. It is suitable for use below ground to carry pressurized water (an archaic use), for garden pots or building decoration in many environments, and for oil containers, oil lamps, or ovens. Most other uses, such as for tableware, sanitary piping, or building decoration in freezing environments, require the material to be glazed. Terracotta, if uncracked, will ring if lightly struck.

Painted ("polychrome") terracotta is typically first covered with a thin coat of gesso, then painted. It has been very widely used but the paint is only suitable for indoor positions and is much less durable than fired colors in or under a ceramic glaze. Terracotta sculpture was very rarely left in its "raw" fired state in the West until the 18th century.

Advantages in Sculpture

As compared to bronze sculpture, terracotta uses a far simpler and quicker process for creating the finished work with much lower material costs. The easier task of modelling, typically with a limited range of knives and wooden shaping tools, but mainly using the fingers, allows the artist to take a more free and flexible approach. Small details that might be impractical to carve in stone, of hair or costume for example, can easily be accomplished in terracotta, and drapery can sometimes be made up of thin sheets of clay that make it much easier to achieve a realistic effect.

Reusable mold-making techniques may be used for production of many identical pieces. Compared to marble sculpture and other stonework the finished product is far lighter and may be further painted and glazed to produce objects with color or durable simulations of metal patina. Robust durable works for outdoor use require greater thickness and so will be heavier, with more care needed in the drying of the unfinished piece to prevent cracking as the material shrinks. Structural considerations are similar to those required for stone sculpture; there is a limit on the stress that can be imposed on terracotta, and terracotta statues of unsupported standing figures are limited to well under life-size unless extra structural support is added. This is also because large figures are extremely difficult to fire, and surviving examples often show sagging or cracks. The Yixian figures were fired in several pieces, and have iron rods inside to hold the structure together.

ARCHITECTURAL TERRACOTTA

Architectural terracotta refers to a fired mixture of clay and water that can be used in a non-structural, semi-structural, or structural capacity on the exterior or interior of a building. Terracotta is an ancient building material. It can be unglazed, painted, slip glazed, or glazed. A piece of terracotta is composed of a hollow clay web enclosing a void space or cell. The cell can be installed in compression with mortar or hung with metal anchors. All cells are partially backfilled with mortar.

Chemistry

Terracotta is made of a clay or silt matrix, a fluxing agent, and grog or bits of previously fired clay. Clays are the remnants of weathered rocks that are smaller than 2 microns. They are composted of silica and alumina. Kaolinite, halloysite, montmorillonite, illite and mica are all good types of clays for ceramic production. When mixed with water they create hydrous aluminum silica that is plastic and moldable. During the firing process the clays lose their water and become a hardened ceramic body.

Fluxes add oxygen when they burn to create more uniform melting of the silica particles throughout the body of the ceramic. This increases the strength of the material. Common fluxing materials are calcium carbonate, alkaline feldspars, manganese, and iron oxides. Grog is used to prevent shrinking and provide structure for the fine clay matrix.

Manufacturing Process

Terracotta can be made by pouring or pressing the mix into a plaster or sandstone mold, clay can be hand carved, or mix can be extruded into a mold using specialized machines. Clay shrinks as it dries from water loss therefore all molds are made slightly larger than the required dimensions. After the desired green-ware, or air dried, shape is created it is fired in a kiln for several days where it shrinks even further. The hot clay is slowly cooled then hand finished. The ceramics are shipped to the project site where they are installed by local contractors. The hollow pieces are partially backfilled with mortar then placed into the wall, suspended from metal anchors, or hung on metal shelf angles.

Design

Academically trained artists were often the designers of the terracotta forms. Their drawings would be interpreted by the manufacturer who would plan out the joint locations and anchoring system. Once finalized, the drawings were turned into a plaster reality by sculptors who would create the mold for the craftsmen.

Vertical pugmill used by the Moravian Pottery and Tile Works in Pennsylvania to refine the clay used for tile production

Clay Preparation

Clay selection was very important to manufacture of terracotta. Homogenous, finer grain sizes were preferred. The color of the clay body was determined by the types of deposits that were locally available to the manufacture. Sand was added to temper the process. Crushed ceramic scraps called grog were also added to stiffen the product and help reduce shrinkage.

Weathering the clay allowed pyrites to chemically change to hydrated ferric oxide and reduced alkali content. This aging minimizes the potential chemical changes during the rest of the manufacturing process. The weathered raw clay was dried, ground, and screened. Later, it would have been pugged in a mill that would mix the clay with water using rotating blades and force the blend through a sieve.

Hand Pressing Terracotta

An artist makes a negative plaster mold based on a clay positive prototype. 1–1 ¼′ of the clay/water mixture is pressed into the mold. Wire mesh or other stiffeners are added to create the web, or clay body that surrounds the hollow cell. The product is air dried to allow the plaster to suck the moisture out of the green clay product. It is fired then slowly cooled.

Extrusion

Mechanized extrusion was used for the mass-production of terracotta blocks, popular in the 1920s. Prepared clay was fed into a machine that would then push the mix through a mold. The technique required the blocks to be made with simple shapes, so this process was often used for flooring, roofing, cladding, and later hollow clay tiles.

A downdraft kiln designed for the Pomona Terra Cotta Manufacturing Company in Guilford County, North Carolina.

Glazing

The last step before firing the greenware was glazing. True glazes are made from various salts but prior to the 1890s most blocks were slip glazed or coated with a watered-down version of the clay mix. Liquefying the clay increased the amount of small silica particles that would be deposited on the surface of the block. These would melt during firing and harden. By 1900 almost all colors could be achieved with the addition of salt glazes. Black or brown were made by adding manganese oxide.

Philadelphia Art Museum's terracotta pediment using polychrome glazing.

Firing

The kiln firing process could take days, up to two weeks. The clay is heated slowly to around 500 °C to sweat off the loose or macroscopic water between the molecules. Then the temperature is increased to close to 900 °C to release the chemically bonded water in gaseous form and the clay particles will begin to melt together or sinter. If the kiln reaches 1000 °C then the clay particles will vitrtify and become glass like. After the maximum temperature was reached then the clay was slowly cooled over a few days. During firing a fireskin is created. A fireskin is the glass like "bread crust" that covers the biscuit or interior body.

Various kilns were used as technology developed and capital was available for investment. Muffle kilns were the most common kiln. They were used as early as 1870. The kilns burned gas, coal, or oil that heated an interior chamber from an exterior chamber. The walls "muffled" the heat so the greenware was not directly exposed to the flames.

Down-draught kilns were also widely used. The interior chamber radiated heat around the terracotta by pulling in hot air from behind an exterior cavity wall. Like the muffle wall, the cavity wall protected the greenware from burning.

Installation

The earliest terracotta elements were laid directly into the masonry but as structural metal became more popular terracotta was suspended by metal anchors. The development of cast and later wrought iron as a structural material was closely linked to the rise of terracotta. Cast iron was first used as columns in the 1820s by William Strickland. Over the course of the 19th century metal became more incorporated into construction but it was not widely used structurally until the late 1890s.

A series of disastrous fires earned terracotta a reputation for being a fireproof, light-weight cladding material that could protect metal from melting. Holes were bored in the hollow blocks in choice locations to allow for metal 'J' or 'Z' hooks to connect the blocks to the load bearing steel frame and/or masonry walls. The metal could be hung vertically or anchored horizontally. Pins, clamps, clips, plates, and a variety of other devices were used to help secure the blocks. The joints would then be mortared and the block would be partially backfilled.

Cracking caused by corroding metal anchors at the First Congregational Church of Long Beach, California.

Degradation

The most common reasons for terracotta to fail are: poor manufacturing, improper installation, weathering, freeze/thaw cycling, and salt formation from atmospheric pollution. The porosity of terracotta greatly impacts its performance. The ability or inability for water and pollutants to enter into the material is directly correlated to its structural capacity. Terracotta is very strong in compression but weak in tension and shear strength. Any anomalous material expanding (ice, salts, incompatible fill material, or corroding metal anchors) inside the clay body will cause it to crack and eventually spall.

Inherent faults can severely impact the performance of the material. Improper molding can cause air pockets to form that increase the rate of deterioration. If the block is not fired or cooled properly than the fireskin will not be uniformly adhered to the substrate and can flake off. Likewise, if a glaze is not fired properly it will crack, flake, and fall off. Discolorations can result from mineral impurities such as pyrites or barium carbonates.

A fair amount of damage comes from clumsy transportation, storage, or installation of the material. If the mortar used around and inside the blocks is too strong then the stress will be translated to the terracotta block which will fail over time. Corroding interior metal anchors expand at a faster rate than the surrounding ceramic body causing it to fail from the inside out. Improper loading of the hollow terracotta blocks can create stress cracks.

The environment also plays a large role in the survival of terracotta. Different types of air pollution can cause different types of surface problems. When it rains, water and salts get sucked into the voids in and around the terracotta through capillary action. If it freezes then ice forms, putting internal stress on the material, causing it to crack from inside. A similar problem happens with atmospheric pollutants that are carried into the gaps by rains water. The pollution creates a mildly acidic solution that eats at the clay body or a salt crust forms, causing similar issues as ice.

ZIRCONIUM DIOXIDE

Zirconium dioxide (ZrO_2), sometimes known as zirconia, is a white crystalline oxide of zirconium. Its most naturally occurring form, with a monoclinic crystalline structure, is the mineral baddeleyite. A dopant stabilized cubic structured zirconia, cubic zirconia, is synthesized in various colours for use as a gemstone and a diamond simulant.

Production, Chemical Properties and Occurrence

Zirconia is produced by calcining zirconium compounds, exploiting its high thermal stability.

Structure

Three phases are known: monoclinic below 1170 °C, tetragonal between 1170 °C and 2370 °C, and cubic above 2370 °C. The trend is for higher symmetry at higher temperatures, as is usually the case. A small percentage of the oxides of calcium or yttrium stabilize in the cubic phase. The very rare mineral tazheranite $(Zr,Ti,Ca)O_2$ is cubic. Unlike TiO_2, which features six-coordinate Ti in all phases, monoclinic zirconia consists of seven-coordinate zirconium centres. This difference is attributed to the larger size of Zr atom relative to the Ti atom.

Chemical Reactions

Zirconia is chemically unreactive. It is slowly attacked by concentrated hydrofluoric acid and sulfuric acid. When heated with carbon, it converts to zirconium carbide. When heated with carbon in the presence of chlorine, it converts to zirconium tetrachloride. This conversion is the basis for the purification of zirconium metal and is analogous to the Kroll process.

Engineering Properties

Zirconium dioxide is one of the most studied ceramic materials. ZrO_2 adopts a monoclinic crystal structure at room temperature and transitions to tetragonal and cubic at higher temperatures. The change of volume caused by the structure transitions from tetragonal to monoclinic to cubic induces large stresses, causing it to crack upon cooling from high temperatures. When the zirconia is blended with some other oxides, the tetragonal and/ or cubic phases are stabilized. Effective dopants include magnesium oxide (MgO), yttrium oxide (Y_2O_3, yttria), calcium oxide (CaO), and cerium(III) oxide (Ce_2O_3).

Zirconia is often more useful in its phase 'stabilized' state. Upon heating, zirconia undergoes disruptive phase changes. By adding small percentages of yttria, these phase changes are eliminated, and the resulting material has superior thermal, mechanical, and electrical properties. In some cases, the tetragonal phase can be metastable. If sufficient quantities of the metastable tetragonal phase is present, then an applied stress, magnified by the stress concentration at a crack tip, can cause the tetragonal phase to convert to monoclinic, with the associated volume expansion. This phase transformation can then put the crack into compression, retarding its growth, and enhancing the fracture toughness. This mechanism is known as transformation toughening, and significantly extends the reliability and lifetime of products made with stabilized zirconia.

The ZrO_2 band gap is dependent on the phase (cubic, tetragonal, monoclinic, or amorphous) and preparation methods, with typical estimates from 5–7 eV.

A special case of zirconia is that of tetragonal zirconia polycrystal, or TZP, which is indicative of polycrystalline zirconia composed of only the metastable tetragonal phase.

Uses

The main use of zirconia is in the production of hard ceramics, such as in dentistry, with other uses including as a protective coating on particles of titanium dioxide pigments, as a refractory material, in insulation, abrasives and enamels. Stabilized zirconia is used in oxygen sensors and fuel cell membranes because it has the ability to allow oxygen ions to move freely through the crystal structure at high temperatures. This high ionic conductivity (and a low electronic conductivity) makes it one of the most useful electroceramics. Zirconium dioxide is also used as the solid electrolyte in electrochromic devices.

Zirconia is a precursor to the electroceramic lead zirconate titanate (*PZT*), which is a high-K dielectric, which is found in myriad components.

Niche uses

The very low thermal conductivity of cubic phase of zirconia also has led to its use as a thermal barrier coating, or TBC, in jet and diesel engines to allow operation at higher temperatures. Thermodynamically, the higher the operation temperature of an engine, the greater the possible efficiency. Another low thermal conductivity use is a ceramic fiber insulation for crystal growth furnaces, fuel cell stack insulation and infrared heating systems.

This material is also used in dentistry in the manufacture of 1) subframes for the construction of dental restorations such as crowns and bridges, which are then veneered with a conventional feldspathic porcelain for aesthetic reasons, or of 2) strong, extremely durable dental prostheses constructed entirely from monolithic zirconia, with limited but constantly improving aesthetics. Zirconia stabilized with yttria (yttrium oxide), known as yttria-stabilized zirconia, can be used as a strong base material in some full ceramic crown restorations.

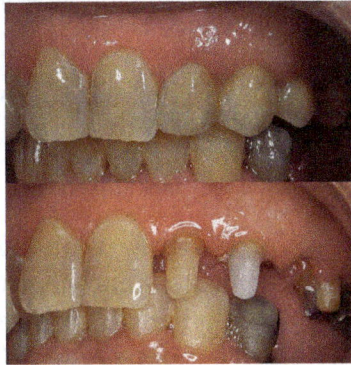

High translucent Zirconia bridge layered by
porcelain and stained with luster paste.

Transformation toughened zirconia is used to make ceramic knives. Because of the hardness, ceramic-edged cutlery stays sharp longer than steel edged products.

Due to its infusibility and brilliant luminosity when incandescent, it was used as an ingredient of sticks for limelight.

Zirconia has been proposed to electrolyze carbon monoxide and oxygen from the atmosphere of Mars to provide both fuel and oxidizer that could be used as a store of chemical energy for use with surface transportation on Mars. Carbon monoxide/oxygen engines have been suggested for early surface transportation use as both carbon monoxide and oxygen can be straightforwardly produced by zirconia electrolysis without requiring use of any of the Martian water resources to obtain hydrogen, which would be needed for the production of methane or any hydrogen-based fuels.

Zirconia is also a potential high-k dielectric material with potential applications as an insulator in transistors.

Zirconia is also employed in the deposition of optical coatings; it is a high-index material usable from the near-UV to the mid-IR, due to its low absorption in this spectral region. In such applications, it is typically deposited by PVD.

In jewelry making, some watch cases are advertised as being "black zirconium oxide". In 2015 Omega released a fully ZrO_2 watch named "The Dark Side of The Moon" with ceramic case, bezel, pushers and clasp, advertising it as four times harder than stainless steel and therefore much more resistant to scratches during everyday use.

Diamond Simulant

Brilliant-cut cubic zirconia.

Single crystals of the cubic phase of zirconia are commonly used as diamond simulant in jewellery. Like diamond, cubic zirconia has a cubic crystal structure and a high index of refraction. Visually discerning a good quality cubic zirconia gem from a diamond is difficult, and most jewellers will have a thermal conductivity tester to identify cubic zirconia by its low thermal conductivity (diamond is a very good thermal conductor). This state of zirconia is commonly called *cubic zirconia*, *CZ*, or *zircon* by jewellers, but the last name is not chemically accurate. Zircon is actually the mineral name for naturally occurring zirconium silicate ($ZrSiO_4$).

COADE STONE

Nelson Pediment, Old Royal Naval College, Greenwich.

Coade stone or Lithodipyra or Lithodipra was stoneware that was often described as an artificial stone in the late 18th and early 19th centuries. It was used for moulding neoclassical statues, architectural decorations and garden ornaments that both were of the highest quality and remain virtually weatherproof today. Produced by appointment to George III and the Prince Regent, it features on St George's Chapel, Windsor; The Royal Pavilion, Brighton; Carlton House, London; the Royal Naval College, Greenwich; and a large quantity was used in the refurbishment of Buckingham Palace in the 1820s.

Lion and Unicorn, entrance to Kensington Palace.

Lithodipyra was first created around 1770 by Eleanor Coade who ran Coade's Artificial Stone Manufactory, Coade and Sealy, and Coade in Lambeth, London, from 1769 until her death in 1821, after which *Lithodipyra* continued to be manufactured by her last business partner, William Croggon, until 1833.

The recipe and techniques for producing Coade stone have been rediscovered by Coade Ltd., who produce sculpture at their workshops in Wilton, Wiltshire.

Home of Eleanor Coade, Belmont House, in Lyme Regis,
Dorset, with Coade stone ornamental façade.

In 1769 Mrs Coade bought Daniel Pincot's struggling artificial stone business at Kings Arms Stairs, Narrow Wall, Lambeth, a site now under the Royal Festival Hall. This business developed into *Coade's Artificial Stone Manufactory* with Eleanor in charge, such that within two years she fired Pincot for 'representing himself as the chief proprietor'.

Mrs Coade did not invent 'artificial stone' – various inferior quality precursors having been both patented and manufactured over the previous forty (or sixty) years – but she was probably responsible for perfecting both the clay recipe and the firing process. It is possible that Pincot's business was a continuation of that run nearby by Richard Holt, who had taken out two patents in 1722 for a kind of liquid metal or stone and another for making china without the use of clay, but there were many start-up 'artificial stone' businesses in the early 18th century of which only Mrs Coade's succeeded.

The company did well, and boasted an illustrious list of customers such as George III and members of the English nobility. In 1799 Mrs Coade appointed her cousin John Sealy (her mother's sister Mary's son), already working as a modeller, as a partner in her business, which then traded as 'Coade and Sealy' until his death in 1813 when it reverted to just 'Coade'.

In 1799 she opened a show room *Coade's Gallery* on *Pedlar's Acre* at the Surrey end of Westminster Bridge Road to display her products.

In 1813 Mrs Coade took on William Croggan from Grampound in Cornwall, a sculptor and distant relative by marriage (second cousin once removed). He managed the factory until her death eight years later in 1821 whereby he bought the factory from the executors for c. £4000. Croggan supplied a lot of Coade stone for Buckingham Palace; however, he went bankrupt in 1833 and died two years later. Trade declined, and production came to an end in the early 1840s.

Modern Replica

In 2000 Coade ltd started producing statues, sculptures and architectural ornament, using the original recipes and methods of the eighteenth century.

Material

(Above) Lion Gate, an entrance into Kew Gardens, with its Coade stone lion statue on top. (Below) The corresponding Coade stone unicorn statue on top of Unicorn Gate, another entrance into the Gardens.

Coade stone is a type of stoneware. Mrs Coade's own name for her products was *Lithodipyra*. Its colours varied from light grey to light yellow (or even beige) and its surface is best described as having a matte finish.

The ease with which the product could be moulded into complex shapes made it ideal for large statues, sculptures and sculptural façades. Moulds were often kept for many years, for repeated use. One-offs were clearly much more expensive to produce, as they had to carry the entire cost of creating the mould.

One of the more striking features of Coade stone is its high resistance to weathering, with the material often faring better than most types of stone in London's harsh environment. Examples of Coade stonework have survived very well; prominent examples are listed below, having survived without apparent wear and tear for 150 years.

Demise

Coade stone was superseded by portland cement as a form of artificial stone and it appears to have been largely phased out by the 1840s.

Quality Controversy

Although Coade stone's reputation for both weather resistance and manufacturing quality is virtually untarnished, three sources describe Rossi's statue of George IV erected in the Royal Crescent, Brighton as "unable to withstand the weathering effects of sea-spray and strong wind: such that, by 1807 the fingers on the sculpture's left hand had been destroyed, and soon afterwards the whole right arm dropped off." By contrast however *Fashionable Brighton, 1820-1860* by Antony Dale describes similar damage as 'wore badly' but does not attribute 'broken fingers, nose, mantle and arm on an unloved statue' to weathering or poor quality Coade stone. In 1819, after considerable complaints, the relic was removed and its present state is undocumented.

A few works produced by Coade, mainly dating from the later period, have shown poor resistance to weathering due to a bad firing in the kiln where the material was not brought up to a sufficient temperature.

Formula

The recipe for Coade stone is still used by Coade Ltd. Rather than being based on cement, it is a ceramic material.

Its manufacture required special skills: extremely careful control and skill in kiln firing, over a period of days. This skill is even more remarkable when the potential variability of kiln temperatures at that time is considered. Coade's factory was the only really successful manufacturer.

The formula used was:

- 10% of grog.

- 5–10% of crushed flint.

- 5–10% fine quartz.

- 10% crushed soda lime glass.

- 60–70% ball clay from Dorset and Devon.

This mixture was also referred to as "fortified clay" which was then inserted after kneading into a kiln which would fire the material at a temperature of 1,100 °C for over four days.

A number of different variations of the recipe were used, depending on the size and fineness of detail in the work a different size and proportion of grog was used. In many pieces a combination of fine grogged Coade clay was used on the surface for detail, backed up by a more heavily grogged mixture for strength.

OPTICAL CERAMICS

Optical ceramics are the advanced industrial materials developed for use in optical applications.

Optical materials derive their utility from their response to infrared, optical, and ultraviolet light. The most obvious optical materials are glasses, but ceramics also have been developed for a number of optical applications.

Passive Devices

Optical and Infrared Windows

In their pure state, most ceramics are wide-band-gap insulators. This means that there is a large gap of forbidden states between the energy of the highest filled electron levels and the energy of the next highest unoccupied level. If this band gap is larger than optical light energies, these ceramics will be optically transparent (although powders and porous compacts of such ceramics will be white and opaque due to light scattering). Two applications of optically transparent ceramics are windows for bar-code readers at supermarkets and infrared radome and laser windows.

Sapphire (a single-crystal form of aluminum oxide, Al_2O_3) has been used for supermarket checkout windows. It combines optical transparency with high scratch resistance. Similarly, single-crystal or infrared-transparent polycrystalline ceramics such as sodium chloride (NaCl), rubidium-doped potassium chloride (KCl), calcium fluoride (CaF), and strontium fluoride (SrF_2) have been used for erosion-resistant infrared radomes, windows for infrared detectors, and infrared laser windows. These polycrystalline halide materials tend to transmit lower wavelengths than oxides, extending down to the infrared region; however, their grain boundaries and porosity scatter radiation. Therefore, they are best used as single crystals. As such, however, halides are insufficiently strong for large windows: they can plastically deform under their own weight. In order to strengthen them, single crystals are typically hot-forged to induce clean grain boundaries and large grain sizes, which do not decrease infrared transmission significantly but allow the body to resist deformation. Alternatively, large-grained material can be fusion-cast.

Lamp Envelopes

Electric discharge lamps, in which enclosed gases are energized by an applied voltage and thereby made to glow, are extremely efficient light sources, but the heat and corrosion involved in their operation push optical ceramics to their thermochemical limits. A major breakthrough occurred in 1961, when Robert Coble of the General Electric Company in the United States demonstrated that alumina (a synthetic polycrystalline, Al_2O_3) could be sintered to optical density and translucency using magnesia (magnesium oxide, MgO) as a sintering aid. This technology permitted the extremely hot sodium discharge in the high-pressure sodium-vapour lamp to be contained in a refractory material that also transmitted its light. The plasma within the inner alumina lamp envelope reaches temperatures of 1,200 °C (2,200 °F). Energy emission covers almost the entire visible spectrum, creating a bright white light that reflects all colours—unlike that of the low-pressure sodium-vapour lamp, whose amber glow is common in the skylines of major cities.

Pigments

The ceramic colour or pigment industry is a long-standing, traditional industry. Ceramic pigments or stains are made of oxide or selenide compounds in combination with specific transition-metal or rare-earth elements. Absorption of certain wavelengths of light by these species imparts specific colours to the compound. For example, cobalt aluminate ($CoAl_2O_4$) and cobalt silicate (Co_2SiO_4) are blue; tin-vanadium oxide (known as V-doped SnO_2) and zirconium-vanadium oxide (V-doped ZrO_2) are yellow; cobalt chromite ($CoCr_2O_3$) and chromium garnet ($2CaO \cdot Cr_2O_3 \cdot 3SiO_2$) are green; and chromium hematite ($CrFe_2O_3$) is black. A true red colour, unavailable in naturally occurring silicate materials, is found in solid solutions of cadmium sulfide and cadmium selenide (CdS-CdSe).

Powdered pigments are incorporated into ceramic bodies or glazes in order to impart colour to the fired ware. Thermal stability and chemical inertness during firing are important considerations.

Active Devices

Phosphors

Ceramic phosphors are employed for both general lighting (as in fluorescent lights) and for electronic imaging (as in cathode-ray tubes). Phosphors function when electrons within them are stimulated from stable, low-energy positions to higher levels by an appropriate means—e.g., thermal, optical, X-ray, or electron excitation. When the energized electrons drop back to lower energy levels, light can be emitted at one or more characteristic wavelengths. These wavelengths are determined by controlled dopants, referred to as activators. Examples of activated phosphors (and their resulting colour emissions) are lead-activated calcium tungstate (blue), manganese-activated zircon (green), lead- or manganese-activated calcium silicate (yellow to orange), and europium-activated yttrium vanadate (red). There are countless other examples.

Two major applications of phosphor ceramics are in cathode-ray tubes (CRTs) for television sets and computer monitors. Thin layers of phosphor powders are applied to the inside of the display screen of the CRT. Electrons are accelerated from the cathode toward the screen, directed by magnetic coils. Light emission (phosphorescence) occurs wherever the electron beam strikes the phosphor layer, and images are formed by high-speed scanning of the electron beam over the surface of the screen. Colour screens employ interspersed small dots of phosphors of each of the three primary colours (red, yellow, and blue), with separate electron beams to address each colour.

Efficient indoor lighting is usually accomplished by fluorescent lamps. Phosphors of a suitably doped calcium halophosphate are deposited as thin powder layers on the inner surfaces of thin-walled glass tubes. The tubes are evacuated and backfilled with a mixture of mercury vapour and an inert gas. An electric discharge through the gas causes the mercury vapour to emit energy in the ultraviolet range, which strikes the phosphor layer and stimulates visible light emission. The resulting combination of blue and orange emission is comparable to that of incandescent lamps.

Phosphors must be manufactured by clean-room methods in order to eliminate unwanted impurities that can "kill" phosphorescence.

Lasers

Lasing, or "light amplification by stimulated emission of radiation," takes place in various media, including glasses and single-crystal ceramics. The first laser, operated by Theodore H. Maiman in 1960, consisted of a rod of synthetic ruby (single-crystal Al_2O_3 doped with chromium) that was excited by a flash lamp. Excitation, or pumping, involves promoting electrons within the dopant centres to higher energy levels by optical or electronic means. The decay of the stimulated electron to a lower energy state yields emission of light, which is contained within the lasing solid between two mirrors (one completely silvered and one partially silvered). As the emitted light reflects back and forth, it stimulates other centres until an intense, coherent, narrow beam of monochromatic light is released. Two well-known ceramic lasing materials are the chromium-doped Al_2O_3 known as ruby and a neodymium-doped yttrium aluminum garnet known as Nd-YAG.

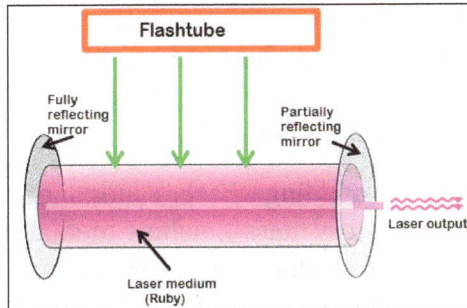

Ruby laser being used in a Q-switch, a special switching device that produces giant output pulse.

Electro-optical Components

Electro-optical ceramics are materials that combine optical transparency with voltage-variable optical, or electro-optical (EO), behaviour. Single-crystal EO materials include lithium niobate ($LiNbO_3$) and lithium tantalate ($LiTaO_3$); polycrystalline EO materials include a lanthanum-modified lead zirconate tantalate known as PLZT. Among other EO properties, these materials exhibit voltage-dependent birefringence. Birefringence is the difference between the refractive index parallel to the optical axis of the crystal and the refractive index perpendicular to the optical axis. Because the propagation velocity is different in the two directions, a phase shift occurs, and this phase shift can be varied by an applied voltage. Such EO behaviour is the basis of a number of optical devices, including switches, modulators, and demodulators for high-speed optical communications. EO ceramic thin films also can be integrated with silicon semiconductors in so-called optoelectronic integrated circuits (OEICs).

Optical ceramics constitute only one of several types of electroceramics.

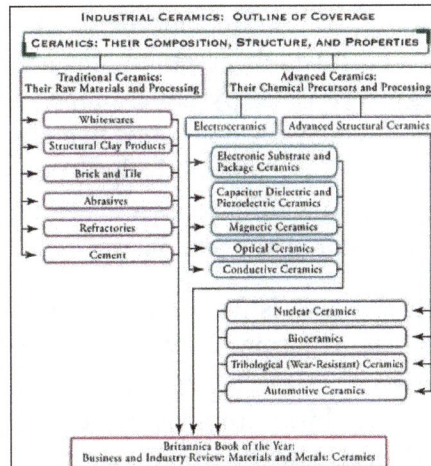

ABRASIVES

Abrasive is a sharp and hard material used to wear away the surface of softer, less resistant materials. Included within the term are both natural and synthetic substances, ranging from the relatively soft particles used in household cleansers and jeweler's polish to the hardest known material, the diamond. Abrasives are indispensable to the manufacture of nearly every product made today.

Abrasives are used in the form of grinding wheels, sandpapers, honing stones, polishes, cutoff wheels, tumbling and vibratory mass-finishing media, sandblasting, pulpstones, ball mills, and still other tools and products. Only through the use of abrasives is industry able to produce the highly precise components and ultrasmooth surfaces required in the manufacture of automobiles, airplanes and space vehicles, mechanical and electrical appliances, and machine tools.

Abrasive Materials: Their Composition and Properties

The materials used to make abrasives can be broadly classified as either natural or synthetic. Natural abrasives include diamond, corundum, and emery; they occur in natural deposits and can be mined and processed for use with little alteration. Synthetic abrasives, on the other hand, are the product of considerable processing of raw materials or chemical precursors; they include silicon carbide, synthetic diamond, and alumina (a synthetic form of corundum). Most natural abrasives have been replaced by synthetic materials because nearly all industrial applications demand consistent properties. With the exception of natural diamond, most of nature's abrasives are too variable in their properties.

One of the most important properties necessary in an abrasive material is hardness. Simply put, the abrasive must be harder than the material it is to grind, polish, or remove. Hardness of the various abrasive materials can be measured on a number of scales, including the Mohs hardness test, the Knoop hardness test, and the Vickers hardness test. The Mohs scale, first described in 1812, measures resistance to indentation as judged by which material will scratch another. This scale, which assigns numbers to natural minerals, has been widely accepted and is used by mineralogists. The Knoop and Vickers hardness tests employ pyramid-shaped diamond indenting devices and measure the indentation made by the diamonds in a given test material. The Vickers test was designed primarily for metals. With the Knoop test, however, the hardness of extremely brittle materials including glass and even diamonds can be measured without harming either the indenter or the test piece.

Toughness or body strength characteristics are also significant to abrasive function. Ideally, a single abrasive particle resharpens itself by the breakdown of its dull cutting or working edge, which exposes another cutting edge within the same particle. In synthetic abrasives it is possible to achieve some degree of control over this property by varying grain shape during the crushing or sizing operation, by making changes in the purity of the abrasive, by alloying abrasives, and by controlling the crystal structure within abrasive grains. Thus abrasives can be developed to meet the operating conditions found in a variety of applications.

Interaction between the abrasive and the material being ground prevents the use of one abrasive as a universal medium. For example, when silicon carbide is used on steel, or alumina on glass, some reaction takes place that has yet to be clearly defined but that results in rapid dulling and inefficient abrasive action. Attrition resistance is the name given to this third, very significant property.

The table lists prominent natural and synthetic abrasive materials.

Hardness of prominent abrasive materials				
Abrasive Materials		Hardness		
		Mohs Scale	Vickers Scale	Knoop Scale
Natural Abrasives	Industrial Diamond	10	10,000	8,000
	Corundum	9	2,200	1,600–2,100
	Emery	7–9	1,600	800–1,800
	Garnet	7–8	1,100–1,300	1,300–1,350
	Flint	7	900–1,100	700–800
	Quartz	7	1,100	700–800
	Pumice	5–6	—	430–560
	Talc	1	—	—
Synthetic Abrasives	Synthetic Diamond	10	10,000	8,000–10,000
	Boron Nitride (Cubic)	10	7,300–10,000	4,700–10,000
	Boron Carbide	9–10	3,300–4,300	2,200–5,100
	Silicon Carbide	9	2,800–3,300	2,000–3,700
	Alumina	9	2,200	2,000–2,600

Fabrication into useful Forms

Preparation and Sizing

All abrasives, with the exception of the naturally appearing fine powders such as talc, must be crushed to the particle size required for use. Sizes in use vary from 4 grit, which measures about 6 millimetres (1/4 inch) in diameter, to as fine as 900 grit, which measures about six microns (0.00024 inch) or about one-tenth the thickness of a human hair. In some cases, even finer powders are required when used for the polishing of scratch-free surfaces on high-quality optical lenses and mirrors for high-power telescopes.

For the coarser sizes the crushed grain is measured by a series of test screens as established in most countries by government standards. For sizes 240 grit and smaller, the size of the grains is usually measured by a settling rate or sedimentation test.

Crushing methods have a significant effect on the cutting action and the strength of abrasive grains. Heavy crushing pressures, for example, tend to create splintery, sharp, and weak grains. These penetrate easily and remove material at a fast rate, and their cutting edges break down easily for resharpening. This shape is required in many coated abrasive products.

When a grinding wheel or coated abrasive belt is to be used to grind away large amounts of material under heavy pressures, a more regularly shaped, strong abrasive grain stands up longer. The sharp edge is not as necessary because heavy pressures are used that effect the penetration of the abrasive into the material being ground.

Abrasive-product Manufacture

Grinding Wheels

The most important abrasive product manufactured is the grinding wheel. Made of abrasive grain and a binder, or "bond," it is a self-sharpening cutting tool. As the grains on the periphery become dull, they are shed from the surface of the grinding wheel, and fresh, sharp cutting edges are exposed. The ability to resharpen is controlled by the nature of the bond used and the ratio of abrasive to bond, measured by volume.

Forming and Firing

In the manufacture of grinding wheels, abrasives of the proper size and bonding materials are weighed and mixed together in a power mixing machine. When thoroughly mixed, measured amounts of the abrasive and bond mixture are evenly distributed in steel molds. The mold is placed in a powerful hydraulic press, and the mixture is compressed to the desired wheel size, allowing some oversize in dimensions for finishing operations. Pressures vary with wheel size and may be upwards of several hundred tons. Some shaving and shaping of the wheels is done before they are baked or fired.

The majority of grinding wheels made have a vitreous, ceramic bond, made of clays and feldspars. The so-called vitrified wheel is fired in high-temperature kilns at temperatures of 1,260 °C (2,300 °F). Electric-, oil-, and gas-fired kilns are used. The length of the "burn" varies with wheel size and can be as long as two weeks.

The remaining 35–40 percent of the grinding wheels manufactured have organic-type bonds using resins, rubber, or shellac as the bonding material. These wheels are baked at temperatures of between 150° and 200 °C (300° and 400 °F). The lower temperature permits the inclusion of steel rings, molded-in threaded bushings, or fibreglass reinforcements, which become baked into the wheel and serve to make it more resistant to breakage from side pressure. The bushings help to hold the wheel in place on certain grinding machines.

Organic wheels can be made much thinner than vitrified wheels, and they are used in place of metal saw blades for cutting a great variety of materials.

Truing, Grading and Testing

Nearly all grinding wheels must be finished after they have been baked or fired. In a process called truing, the wheels are cut to final size, and the outside glazed layers resulting from the kiln are removed, making the sides of the wheel parallel and the size of the arbor hole accurate; at the same time the working surface of the wheel is sharpened. Wheels are trued by using conical steel cutters, by rubbing in beds of steel shot, and by grinding with grinding wheels.

Grading of wheels assures that they have the correct resistance to wear. Grade or hardness, determined by the amount of bond, permits the grinding wheel to keep itself sharp and free-cutting in a variety of conditions. Grinding wheels used for heavy grinding operations are "harder" and are made with greater amounts of bond, retaining the abrasive particles longer under severe conditions such as those found in steel mills and foundries. In the tool room where industrial cutting tools are sharpened, softer wheels, with less bond per unit of abrasive, are required, so that, as soon as the abrasive grains start to dull and the possibility develops of building up heat in the sensitive cutting tool being ground, the wheel will resharpen and shed the dulled grains.

Finally, grinding wheels are checked for balance to assure that they will run without vibration. Grinding wheels six inches in diameter and larger are usually speed-tested. The wheel is rotated at a speed at least 50 percent greater than the maximum allowable operating speed. This is a nondestructive measure of the wheel's bursting strength.

The same basic processes are used in the manufacture of abrasive bricks, sticks, and formed shapes. These are used as rubbing blocks, sharpening stones, honing stones, and shaped abrasive mediums for tumbling or mass finishing.

Sandpapers

Sandpapers (coated abrasive) are the next most significant abrasive product. They consist, basically, of a single layer of abrasive particles held to a flexible backing material by an adhesive bond. The cutting action of coated abrasive products is determined by the abrasive used, the grit size, the density or spacing of the grit, the strength of the adhesive, and the flexibility of the backing material.

Coating

Manufacture begins with huge rolls of backing material, either paper, cloth, or a combination of the two. The backing is fed to the making machine where the first layer of adhesive is applied. Next, the layer of abrasive is applied, either by gravity or electrostatically. The electrostatic method orients the slivery type of abrasive used, with the sharp ends facing out. The process can also control the spacing of the grains. This is an advantage in that wide spacing helps to alleviate loading problems when grinding soft, stringy, or gummy materials.

After being coated with the abrasive layer, the product is draped in long festoons in which it partially dries. Then it is run through another sizing operation and a second layer of adhesive is applied. The product is draped, allowed to dry thoroughly, and wound into large-diameter rolls.

The adhesives used to bond the abrasive to the backing are water-soluble, water-proof, or a combination of the two. Water-soluble types are used for dry grinding operations and in household sanding, and occasionally on wood-sanding commercial operations. Resin or resin-over-glue types currently in use have the flexibility associated with the soluble types. The all-resin type is best for severe operations; its properties are such that the heat of grinding actually increases the adhesive's holding strength.

Shaping

Sandpaper disks for right-angle grinders are die-cut. Sheets are cut in the standard lengths and widths used in production and household applications. Strips are slit, cut to length, and joined by their ends to make up the coated abrasive belts that have become an essential part of industry, replacing several kinds of abrasive wheels used previously.

Other Abrasive Products

Other products use the abrasive in the form of grains or powders. In addition to the sizing operation, many types are specifically treated, by calcining, acid, or heating, to make them more suitable for use as lapping abrasive or perhaps as sandblasting grain. For use in lapping and polishing, the abrasive is usually mixed with a vehicle such as mineral or seal oil. Polishing sticks consist of waxes or greases impregnated with various-sized abrasive grains, depending on the particular requirements of the work.

Two materials used for cleaning rather than grinding are still abrasive in nature. Glass beads, pressure blasted onto a surface, remove rust, scale, and carbon. These have replaced much hand cleaning with steel wool. Steel wool still has some applications.

Industrial Applications

Grinding

Grinding, the most important abrasive application, is in some way involved in the manufacture of almost every product. This use may be direct, as when the product requires pieces that must be made within close dimensional tolerance limits, or a very smooth surface, or when used on materials too hard to be machined by conventional cutting tools; or indirect, as when, for example, grinding wheels are used to sharpen cutting tools. The materials that are used to make cutting tools must of course be hard in order for the cutting tool to cut and retain its sharp edge. Abrasive grinding wheels are the only means for sharpening the dull edges of such tools.

Grinding wheels in use in industry today rotate at peripheral speeds of almost 300 kilometres (180 miles) per hour. The abrasive wheel may throw a long stream of bright yellow sparks and remove upwards of half a ton of metal per hour while grinding the imperfections from the surface of a bar of stainless steel. Or the grinding wheel may be as small as 0.55 millimetre (0.022 inch) in diameter, may rotate at 150,000 revolutions per minute, and may grind miniature precision ball bearings to accuracies measured in micrometres.

In the automotive industry, only abrasives can produce the tight fit required between piston rings and cylinders to prevent the escape of compressed gasoline vapours. Valves and valve seats are ground. Bearing surfaces in the engine, transmission, and wheels need specific finish, size, and roundness to assure frictionless rotation. These can be achieved only with abrasive tools.

Abrasive machining, the use of abrasives rather than high-speed steel or tungsten carbide cutting tools, makes use of the self-sharpening grinding wheel and eliminates tool sharpening costs. The ability to grind hardened materials without the previously necessary prehardening machining saves intermediate part-handling operations.

Speed and improved grinding systems, machines, and grinding tools are the main reason for the increased importance of grinding. One-hundred-horsepower motors, automatic loading equipment, high-speed grinding wheels removing large amounts of hard-to-grind materials, ultrafine tolerances, and costly machines are part of the new abrasives systems that are capable of extremely high rates of production. Yet the abrasive products portion of the total cost to grind the part may be as low as 5 percent, even though the grinding wheel, with diamond as the abrasive, may cost thousands of dollars. Or the wheel may cost a few pennies and be used to shape a die used in the manufacture of tableware.

Tumbling Media

Relatively new products, such as plastic bonded tumbling media for mass finishing a multitude of parts, have eliminated time-consuming hand-deburring operations, a plague to the aircraft industry, in which the high cost of labour makes handwork prohibitively expensive. Parts with rough edges are tumbled in a rotating barrel of loose abrasive or preshaped abrasive pieces. As the mass slides, burrs are ground away, surfaces are finished, and edges are smoothed. When the same parts and media are mixed in a vibrating tub, the process becomes even more productive.

Cutting Wheels

Abrasive wheels have replaced steel saws in many places. Thin, abrasive cutoff wheels are capable of sawing through nearly every material known, at rates faster than those of metal saws, while generating less heat and producing a better cut surface. Some space-age metals, because of their hardnesses, can be cut only with abrasive wheels. Granite, marble, slate, and various building blocks are cut to size with diamond abrasive wheels. Grooves for expansion joints and for the reduction of wet-weather skidding accidents are cut in concrete runways and highways by blades with a metal centre, onto which are brazed metal segments with the diamond abrasive mixed throughout.

Tool Sharpening

The sharpening of all types of tools continues to be a major grinding operation. Drills, saws, reamers, milling cutters, broaches, and the great spectrum of knives are kept sharp by abrasives. Coarser-grit products are used for their initial shaping. Finer-grit abrasives produce keener cutting edges. Ultrasharp tools must be hand-honed on natural sharpening or honing stones. Even grinding wheels themselves may require some sharpening. Specially designed steel disks or diamond tools are used to remove dull abrasive cutting edges and create a sharp cutting surface.

Metal Cleaning

In foundries and steel mills, grinding wheels and coated abrasive belts remove the unwanted portions of castings, forgings, and billets. Abrasive grit is pressure-blasted against the metal to clean it in preparation for painting. Metal shot is used on softer metallic castings.

Miscellaneous Applications

The roster of unusual applications for abrasives includes cutting frozen fish into fish sticks; grinding animal-gut sutures and guitar strings to constant diameters; removing human skin blemishes and birthmarks or shaping bones by plastic surgery; removing spots and discolorations from suede clothing; grinding toothpicks round; and making stairs skidproof by abrasive grain.

References

- Clay-products-3665: brainkart.com, Retrieved 10 July, 2019

- Haynes, William M., ed. (2011). CRC Handbook of Chemistry and Physics (92nd ed.). Boca Raton, FL: CRC Press. P. 4.88. ISBN 1439855110

- Riley, Frank L. (2004). "Silicon Nitride and Related Materials". Journal of the American Ceramic Society. 83 (2): 245–265. Doi:10.1111/j.1151-2916.2000.tb01182.x

- Refractory, technology: britannica.com, Retrieved 19 April, 2019

- Integral Textile Ceramic Structures Annual Review of Materials Research Vol. 38: 425–443 (Volume publication date August 2008) First published online as a Review in Advance on March 26, 2008 doi:10.1146/annurev.matsci.38.060407.130214

- Ceramic-cements: ceramicindustry.com, Retrieved 25 February, 2019

- Ikesue, A (2002). "Polycrystalline Nd:YAG ceramics lasers". Optical Materials. 19: 183. Bibcode:2002optma..19..183I. Doi:10.1016/S0925-3467(01)00217-8

- Optical-ceramics, technology: britannica.com, Retrieved 21 May, 2019

- Hecht, Jeff (2005). Beam: The Race to Make the Laser. Oxford University Press. ISBN 0-19-514210-1

- Abrasive, technology: britannica.com, Retrieved 8 January, 2019

3

Ceramic Production Techniques

There are numerous techniques which are used to produce ceramics such as sintering, slip casting, freeze-casting, compaction of ceramic powders, ceramic mold casting, ceramic matrix composite and ceramic shell molding. This chapter discusses in detail these production techniques related to ceramics.

CERAMIC FORMING TECHNIQUES

Having a strong background by being the first technique that the human being has learned, ceramic is still applied today with similar techniques used ages ago. Ceramic products differ from each other both in terms of the mixture of the mud and the baking method.

All kinds of mud consist of clay in different qualities (the element that gives form) and the oil removing elements (soil, quartz, ash, pieces of plants, straw, grinded calcite etc.). By adding a melting substance makes the mud glassy and thus it transforms into the porcelain. Generally, the 'ceramic mud' material has various types. These kinds of mud are all different from each other in terms of their chemical qualities, patterns and colors; however, their common characteristic is the feature of being shaped easily and artificially. The thing that qualifies the mud as good is the ability of keeping glaze and becoming usable, durable, and glassy.

Initially, the ceramic product is designed on paper. Then, the design is brought into 3D by choosing the appropriate mud (casting, red clay etc.). In that process, the designer is supposed to choose the most convenient forming method by being aware of that different formations come out by applying different degrees of heat. The final product needs to be the one which forms a whole with its production method, its shape, decoration and glaze. That's why, trying to solve all the details on paper without having the knowledge about the characteristics of the material, may result only in waste of time and material.

In product design, aesthetical values should not be ignored. These values may differ from each other depending on the purposes of use. For instance, tableware products should have the forms that the user can personalize and organize in a way he/she wants. Each pieces of sets like dishes and salt containers should be considered as a part of all the other pieces of the sets like handles and shaped in order to function truly and form an union in terms of material and form.

As already told, while bringing the ceramic into 3D, the most convenient forming method should be chosen. There are many factors that play important roles on choosing the forming method. For example, the structure of the mud, its aim and the field of use, the efficiency of the production, the formal structure of the product needs to be known. By combining the techniques of hand-shaping such as making stripes, preparing mold and making plate, it is possible to form products. While giving shape by hand, mud is kneaded well to make it mix with the mixture and to make the air flow out. The presence of the air may result in some cracks while the product is left for drying or is placed in the furnace. If the kneaded mud will be given shape by 'plating technique', by the help of a roller-pin or wooden molds, a number of plates in different thickness and similar dimensions with the product can be formed. Because the thickness of the plate forms wall thickness of the product, it shouldn't be too thick. The tools that are used to give the shape of mud are called 'ebejuar'. Sticking two different surfaces needs some notches on both sides of the surfaces that should also get wet in order to be taped.

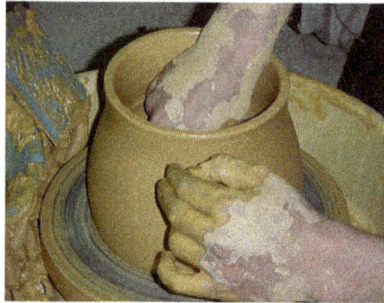

The sticking details are very important while they are applied on mud. Surfaces that are not stuck well may crack during the drying process. Beside the plating technique, there is another technique called 'striping technique' in which the hand made stripes are stuck together on a floor of mud. Another method of giving shape by hand is 'lathe'. Lathe is used when symmetric and round shapes are designed. In lathe technique, it needs to be paid attention to prevent the mud from getting away from the center in order to transform the mud into its form in a balanced way. Lathe is a technique which develops only with practicing. Ceramic mud can also be used in forming wall-panels, sculptures and mass models of some products by taking or adding some pieces. However, after building a mass model, the model needs to be emptied to a determined wall thickness.

When ceramic forming techniques are considered in terms of industry, it is observed that there are some techniques like pressing or casting into a plaster mold (the reason for the usage of plaster is the plaster's ability to assume the water in the mud and to make the mud stick on the surface and dry to form wall-thickness) that are based on mass production. If the product will be given a shape by casting into a plaster mold, technical drawings of the product should be drawn by considering the mud's shrinking ratio and the model should be prepared with plaster lathe. The pins, which make the plaster pieces connect to each other, are placed between the pieces. While making a plaster mold, the details for preparing plaster are also very important. Molds are shaped according to the type of casting (empty or full casting). Large forms like service plates and trays are formed by full casting while empty forms like cups are formed by empty casting. Another method in plaster molds is to press dry mud into the mold by automatic template lathe.

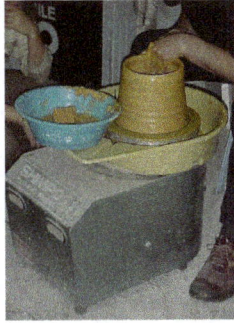

Before being placed into the furnace, the shaped product needs to be fully dried. Drying should be progressed slowly according to the products details, otherwise the product may crack during the drying process or when it is left in the furnace to be baked. The dried surface product is finally applied by sandpaper or foam rubber in order to form a smooth surface and placed in the furnace. The first part of baking must be at least 950 degree Celsius furnaces. To make the baked product durable, glaze is applied with an appropriate technique (plunging, spraying, pouring, dusting etc.). The glazed product is baked for the second time at a high temperature. Temperature of the furnace differs according to the type of the glaze. Glazing the product adds a nice visual effect and a hygienic quality to the product. Having opaque and shiny types, glazes can be varied by adding color.

A ceramic product is applied the same procedures (primary product-drying-baking-glazing-baking) even if the types of their mud and production methods differ from each other. If decoration will be applied on the product, under-glaze and over-glaze paints are used. In the end, a successful process creates a functional and high-quality product. That's also an important factor for both, the manufacturer and the user.

SINTERING

Sintering (Firing) of ceramic materials is the method involving consolidation of ceramic powder particles by heating the "green" compact part to a high temperature below the melting point, when the material of the separate particles difuse to the neghbouring powder particles.

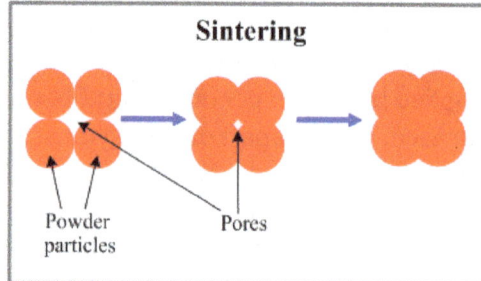

The driving force of sintering process is reduction of surface energy of the particles caused by decreasing their vapour-solid interfaces.

During the diffusion process the pores, taking place in the "green compact", diminish or even close up, resulting in densification of the part, improvement of its mechanical properties.

Decrease of the porosity, caused by the sintering process, is determined by the level of the initial porosity of the "green" compact, sintering temperature and time. Sintering is enhanced if a liquid phase takes part in the process (liquid phase sintering).

Sintering (firing) of pure oxide ceramics require relatively long time and high temperature because the diffusion proceeds in solid state.

Applying pressure decreases sintering time and the resulted porosity.

Tunnel kilns and periodic kilns are commonly used for ceramics sintering (firing). In periodic kilns heating and cooling sintering stages are conducted according to a prescribed procedure. In tunnel kilns the sintered parts are conveyed through different temperature zones.

Typical tunnel kiln has three zones:

- Preheat zone for removing lubricant and other organic materials.
- Sintering zone where the diffusion occurs.
- Cooling zone where the sintered parts cool down.

Sintering process may be conducted in different atmospheres: air, inert atmosphere.

Sintering occurs simultaneously with pressing in the hot pressing processes (hot die pressing, hot isostatic pressing).

SLIP CASTING

Slipcasting or slip casting is a ceramic forming technique for the mass-production of pottery and other ceramics, especially for shapes not easily made on a wheel. Slipcast ware should not be confused with slipware, which is pottery formed by any technique that is decorated using slip. In slipcasting, a liquid clay body slip (usually mixed in a blunger) is poured into plaster moulds and allowed to form a layer, the cast, on the inside walls of the mould.

In a solid cast mould, ceramic objects such as handles and plates are surrounded by plaster on all sides with a reservoir for slip, and are removed when the solid piece is held within. For a hollow cast mould, for objects such as vases and cups, once the plaster has absorbed most of the liquid from the outside layer of clay the remaining slip is poured off for later use. After a period for further absorption of water, the cast piece is removed from the mould once it is leather-hard, that is, firm enough to handle without losing its shape. It is then "fettled" (trimmed neatly) and allowed to dry out further, usually overnight or for several hours. This produces a *greenware* piece which is then ready to be decorated, glazed and fired in a kiln.

The technique is suited to the production of complex shapes, especially if with relief decoration and thin walls. Much modern fine factory porcelain is made by the technique, very often the entire production. It is also commonly used for sanitaryware, such as toilets and basins, and smaller pieces like figurines and teapots. The technique can also be used for small-scale production runs or to produce limited edition, one off objects, especially reproductions of antique dolls and modern porcelain doll-making.

An additive with deflocculant properties, such as sodium silicate, can be added to the slip to disperse the raw material particles. This allows a higher solid content to be used, or allows a fluid slip to be produced with a minimum of water so that drying shrinkage is minimised, which is important during slipcasting.

The French for slip is barbotine ("Coulée en barbotine" is slipcasting), and "barbotine pottery" is sometimes used for 19th century French and American pottery with added slipcast decoration, as well as (confusingly) being the English term for a variety of slip-ware that is decorated with thick blobs of slip.

FREEZE-CASTING

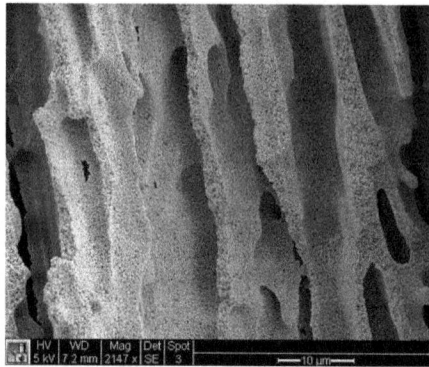

Freeze-Cast alumina that has been partially sintered.
The freezing direction in the image is up.

Freeze-casting, also frequently referred to as *ice-templating*, is a technique that exploits the highly anisotropic solidification behavior of a solvent (generally water) in a well-dispersed slurry to template controllably a directionally porous ceramic. By subjecting an aqueous slurry to a directional temperature gradient, ice crystals will nucleate on one side of the slurry and grow along the temperature gradient. The ice crystals will redistribute the suspended ceramic particles as they grow within the slurry, effectively templating the ceramic.

Once solidification has ended, the frozen, templated ceramic is placed into a freeze-dryer to remove the ice crystals. The resulting green body contains anisotropic macropores in a replica of the sublimated ice crystals and micropores found between the ceramic particles in the walls. This structure is often sintered to consolidate the particulate walls and provide strength to the porous material. The porosity left by the sublimation of solvent crystals is typically between 2–200 μm.

The first observation of cellular structures resulting from the freezing of water goes back over a century, but the first reported instance of freeze-casting, in the modern sense, was in 1954 when Maxwell et al. attempted to fabricate turbosupercharger blades out of refractory powders. They froze extremely thick slips of titanium carbide, producing near-net-shape castings that were easy to sinter and machine. The goal of this work, however, was to make dense ceramics. It was not until 2001, when Fukasawa et al. created directionally porous alumina castings, that the idea of using freeze-casting as a means of creating novel porous structures really took hold. Since that time, research has grown considerably with hundreds of papers coming out within the last decade.

Because freeze-casting is a physical process, the techniques developed for one material system can be applied to a wide range of materials. Additionally, due to the inordinate amount of control and broad range of possible porous microstructures that freeze-casting can produce, the technique has found its niche in a number of disparate fields such as tissue scaffolds, photonics, metal-matrix composites, dentistry, materials science, and even food science

There are three possible end results to uni-directionally freezing a suspension of particles. First, the ice-growth proceeds as a planar front, pushing particles in front like a bulldozer pushes a pile of rocks. This scenario usually occurs at very low solidification velocities ($< 1~\mu m~s^{-1}$) or with extremely fine particles because they can move by Brownian motion away from the front. The resultant structure contains no macroporosity. If one were to increase the solidification speed, the size of the particles or solid loading moderately, the particles begin to interact in a meaningful way with the approaching ice front. The result is typically a lamellar or cellular templated structure whose exact morphology depends on the particular conditions of the system. It is this type of solidification that is targeted for porous materials made by freeze-casting. The third possibility for a freeze-cast structure occurs when particles are given insufficient time to segregate from the suspension, resulting in complete encapsulation of the particles within the ice front. This occurs when the freezing rates are rapid, particle size becomes sufficiently large, or when the solids loading is high enough to hinder particle motion. To ensure templating, the particles must be ejected from the oncoming front. Energetically speaking, this will occur if there is an overall increase in free energy if the particle were to be engulfed $(\Delta\sigma > 0)$.

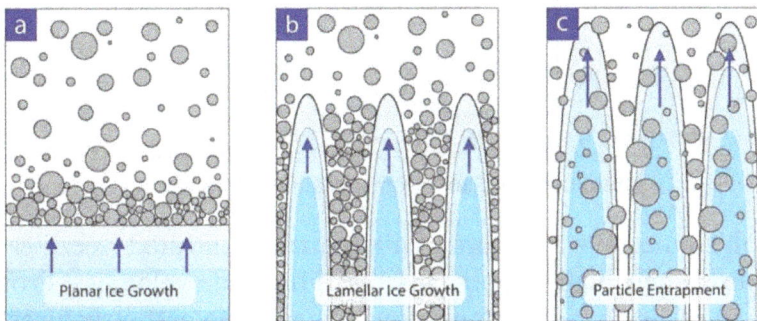

Depending on the speed of the freezing front, particle size and solids loading there are three possible morphological outcomes: (a) planar front where all particles are pushed ahead of the ice, (b) lamellar/cellular front where ice crystals template particles or (c) particles are engulfed producing no ordering.

$$\Delta\sigma = \sigma_{ps} - (\sigma_{pl} + \sigma_{sl})$$

where $\Delta\sigma$ is the change in free energy of the particle, σ_{ps} is the surface potential between the particle and interface, σ_{pl} is the potential between the particle and the liquid phase and σ_{sl} is the surface potential between the solid and liquid phases. This expression is valid at low solidification velocities, when the system is shifted only slightly from equilibrium. At high solidification velocities, kinetics must also be taken into consideration. There will be a liquid film between the front and particle to maintain constant transport of the molecules which are incorporated into the growing crystal. When the front velocity increases, this film thickness (d) will decrease due to increasing drag forces. A critical velocity (v_c) occurs when the film is no longer thick enough to supply the needed molecular supply. At this speed the particle will be engulfed. Most authors express v_c as a function of particle size where . The transition from a porous R (lamellar) morphology to one where the majority of particles are entrapped occurs at v_c, which was defined by Deville et al. to be:

$$v_c = \frac{\Delta\sigma d}{3\eta R}\left(\frac{a_0}{d}\right)^z$$

where a_0 is the average intermolecular distance of the molecule that is freezing within the liquid, d is the overall thickness of the liquid film, η is the solution viscosity, R is the particle radius and z is an exponent that can vary from 1 to 5. As expected, we see that v_c decreases as particle radius R goes up.

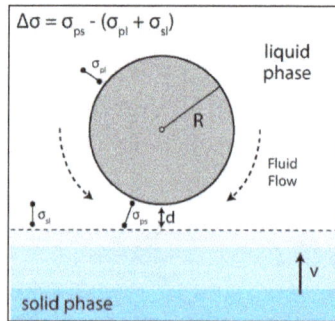

Schematic of a particle within the liquid phase
interacting with an oncoming solidification front.

Waschkies et al. studied the structure of dilute to concentrated freeze-casts from low ($< 1\ \mu m\ s^{-1}$) to extremely high ($> 700\ \mu m\ s^{-1}$) solidification velocities. From this study, they were able to generate morphological maps for freeze-cast structures made under various conditions. Maps such as these are excellent for showing general trends, but they are quite specific to the materials system from which they were derived. For most applications where freeze-casts will be used after freezing, binders are needed to supply strength in the green state. The addition of binder can significantly alter the chemistry within the frozen environment, depressing the freezing point and hampering particle motion leading to particle entrapment at speeds far below the predicted v_c. Assuming, however, that we are operating at speeds below v_c and above those which produce a planar front, we will achieve some cellular structure with both ice-crystals and walls composed of packed ceramic particles. The morphology of this structure is tied to some variables, but the most influential is the temperature gradient as a function of time and distance along the freezing direction.

Freeze-casts have at least three apparent morphological regions. At the side where freezing initiates is a nearly isotropic region with no visible macropores dubbed the Initial Zone (IZ). Directly after the IZ is the Transition Zone (TZ), where macropores begin to form and align with one another. The pores in this region may appear randomly oriented. The third zone is called the Steady-State Zone (SSZ), macropores in this region are aligned with one another and grow in a regular fashion. Within the SSZ, the structure is defined by a value λ that is the average thickness of a ceramic wall and its adjacent macropore.

Initial Zone: Nucleation and Growth Mechanisms

Although the ability of ice to exclude suspended particles has long been known, the mechanism is still being debated. It was believed initially that during the moments immediately following the nucleation of the ice crystals, particles were ejected from the growing planar ice front, leading to the formation of a constitutionally super-cooled zone directly ahead of the growing ice. This unstable region eventually resulted in perturbations, breaking the planar front into a columnar ice front, a phenomenon better known as a Mullins-Serkerka instability. After the breakdown, the ice crystals grow along the temperature gradient, pushing ceramic particles from the liquid phase aside so that they accumulate between the growing ice crystals. However, recent in-situ X-ray radiography of directionally frozen alumina suspensions reveal a different mechanism.

In-situ testing reveals that freeze-casting is an aggressive growth process. In the moments immediately before nucleation, the suspension is in an unstable super-cooled state. Homogeneous (spatially speaking) nucleation of ice crystals occurs followed by explosive crystal growth in every spatial and crystallographic direction. The initial nucleation and growth steps are so rapid (approaching 800 mm s^{-1}) that all suspended particles are completely engulfed by the oncoming ice front because not enough time is given for particle redistribution, resulting in a structure with anisotropic particle distribution. This step is what provides the initial zone structure.

Transition Zone: A Changing Microstructure

As solidification slows and growth kinetics become rate-limiting, the ice crystals begin to exclude the particles, redistributing them within the suspension. A competitive growth process develops between two crystal populations, those with their basal planes aligned with the thermal gradient (z-crystals) and those that are randomly oriented (r-crystals) giving rise to the start of the TZ.

There are colonies of similarly aligned ice crystals growing throughout the suspension. There are fine lamellae of aligned z-crystals growing with their basal planes aligned with the thermal gradient. The r-crystals appear in this cross-section as platelets but in actuality, they are most similar to columnar dendritic crystals cut along a bias. Within the transition zone, the r-crystals either stop growing or turn into z-crystals that eventually become the predominant orientation, and lead to steady-state growth. There are some reasons why this occurs. For one, during freezing, the growing crystals tend to align with the temperature gradient, as this is the lowest energy configuration and thermodynamically preferential. Aligned growth, however, can mean two different things. Assuming the temperature gradient is vertical, the growing crystal will either be parallel (z-crystal) or perpendicular (r-crystal) to this gradient. A crystal that lays horizontally can still grow in line with the temperature gradient, but it will mean growing on its face rather than its edge. Since the thermal conductivity of ice is so small (1.6 - 2.4 W mK^{-1}) compared with most every other ceramic (ex. Al_2O_3 = 40 W mK^{-1}), the growing ice will have a significant insulative effect on the localized thermal conditions within the slurry. This can be illustrated using simple resistor elements.

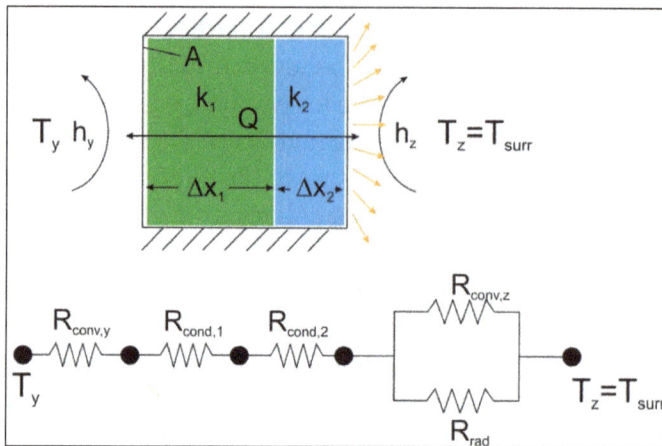

Thermal Resistance.

When ice crystals are aligned with their basal planes parallel to the temperature gradient (z-crystals), they can be represented as two resistors in parallel. The thermal resistance of the ceramic is significantly smaller than that of the ice however, so the apparent resistance can be expressed as the lower $R_{ceramic}$. If the ice crystals are aligned perpendicular to the temperature gradient (r-crystals), they can be approximated as two resistor elements in series. For this case, the R_{ice} is limiting and will dictate the localized thermal conditions. The lower thermal resistance for the z-crystal case leads to lower temperatures and greater heat flux at the growing crystals tips, driving further growth in this direction while, at the same time, the large R_{ice} value hinders the growth of the r-crystals. Each ice crystal growing within the slurry will be some combination of these two scenarios. Thermodynamics dictate that all crystals will tend to align with the preferential temperature gradient causing r-crystals to eventually give way to z-crystals, which can be seen from the following radiographs taken within the TZ.

When z-crystals become the only significant crystal orientation present, the ice-front grows in a steady-state manner except there are no significant changes to the system conditions. It was observed in 2012 that, in the initial moments of freezing, there are dendritic r-crystals that grow 5 - 15 times faster than the solidifying front. These shoot up into the suspension ahead of the main ice front and partially melt back. These crystals stop growing at the point where the TZ will eventually fully transition to the SSZ. Researchers determined that this particular point marks the position where the suspension is in an equilibrium state (i.e. freezing temperature and suspension temperature are equal). We can say then that the size of the initial and transition zones are controlled by the extent of supercooling beyond the already low freezing temperature. If the freeze-casting setup is controlled so that nucleation is favored at only small supercooling, then the TZ will give way to the SSZ sooner.

Steady-state Growth Zone

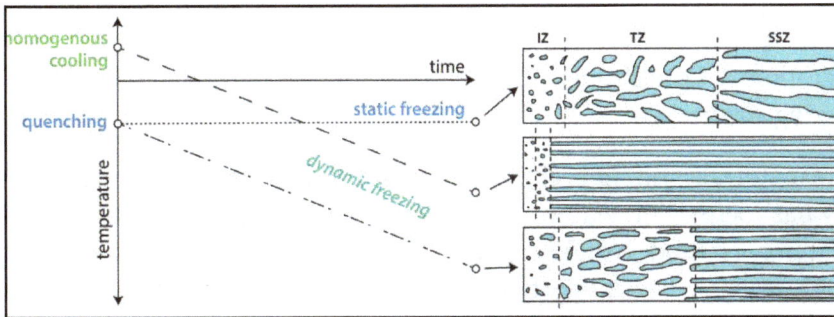

Thermal profiles and their effect on
subsequent microstructure of freeze-casts.

The structure in this final region contains long, aligned lamellae that alternate between ice crystals and ceramic walls. The faster a sample is frozen, the finer its solvent crystals (and its eventual macroporosity) will be. Within the SSZ, the normal speeds which are usable for colloidal templating are 10 – 100 mm s^{-1} leading to solvent crystals typically between 2 mm and 200 mm. Subsequent sublimation of the ice within the SSZ yields a green ceramic preform with porosity in a nearly exact replica of these ice crystals. The microstructure of a freeze-cast within the SSZ is defined by its wavelength (λ) which is the average thickness of a single ceramic wall plus its adjacent macropore. Several publications have reported the effects of solidification kinetics on the microstructures of freeze-cast materials. It has been shown that λ follows an empirical power-law relationship with solidification velocity (v):

$$\lambda = Av^{-n}$$

Both A and v are used as fitting parameters as currently there is no way of calculating them from first principles, although it is generally believed that A is related to slurry parameters like viscosity and solid loading while n is influenced by particle characteristics.

Controlling the Porous Structure

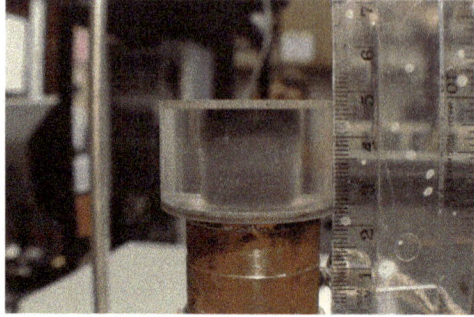

Stop-motion animation of the freeze-casting process.

There are two general categories of tools for architecture a freeze-cast:

1. Chemistry of the System: Freezing medium and chosen particulate material(s), any additional binders, dispersants or additives.

2. Operational Conditions: Temperature profile, atmosphere, mold material, freezing surface, etc.

Initially, the materials system is chosen based on what sort of final structure is needed. This review has focused on water as the vehicle for freezing, but there are some other solvents that may be used. Notably, camphene, which is an organic solvent that is waxy at room temperature. Freezing of this solution produces highly branched dendritic crystals. Once the materials system is settled on however, the majority of microstructural control comes from external operational conditions such as mold material and temperature gradient.

Controlling Pore Size

The microstructural wavelength (average pore + wall thickness) can be described as a function of the solidification velocity v ($\lambda = Av^{-n}$) where A is dependent on solids loading. There are two ways then that the pore size can be controlled. The first is to change the solidification speed that then alters the microstructural wavelength, or the solids loading can be changed. In doing so, the ratio of pore size to wall size is changed. It is often more prudent to alter the solidification velocity seeing as a minimum solid loading is usually desired. Since microstructural size (λ) is inversely related to the velocity of the freezing front, faster speeds lead to finer structures, while slower speeds produce a coarse microstructure. Controlling the solidification velocity is, therefore, crucial to being able to control the microstructure.

Controlling Pore Shape

Additives can prove highly useful and versatile in changing the morphology of pores. These work by affecting the growth kinetics and microstructure of the ice in addition to the topology of the ice-water interface. Some additives work by altering the phase diagram of the solvent. For example, water and NaCl have a eutectic phase diagram. When NaCl is added into a freeze-casting suspension, the solid ice phase and liquid regions are separated by a zone where both solids and liquids can coexist. This briny region is removed during sublimation, but its existence has a strong effect on the microstructure of the porous ceramic. Other additives work by either altering the interfacial surface energies between the solid/liquid and particle/liquid, changing the viscosity of the suspension, or the degree of undercooling in the system. Studies have been done with glycerol, sucrose, ethanol, Coca-Cola, acetic acid and more.

Static vs. Dynamic Freezing Profiles

If a freeze casting setup with a constant temperature on either side of the freezing system is used, (static freeze-casting) the front solidification velocity in the SSZ will decrease over time due to the increasing thermal buffer caused by the growing ice front. When this occurs, more time is given for the anisotropic ice crystals to grow perpendicularly to the freezing direction (c-axis) resulting in a structure with ice lamellae that increase in thickness along the length of the sample.

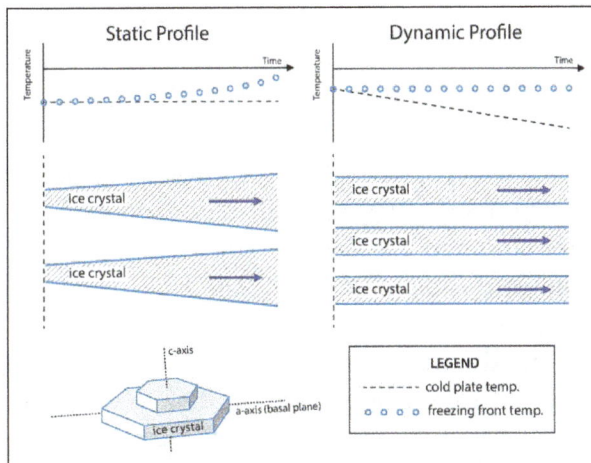

Static and dynamic freezing profiles
in the steady-state freezing regime.

To ensure highly anisotropic, yet predictable solidification behavior within the SSZ, dynamic freezing patterns are preferred. Using dynamic freezing, the velocity of the solidification front, and, therefore, the ice crystal size, can be controlled with a changing temperature gradient. The increasing thermal gradient counters the effect of the growing thermal buffer imposed by the growing ice front. It has been shown that a linearly decreasing temperature on one side of a freeze-cast will result in near-constant solidification velocity, yielding ice crystals with an almost constant thickness along the SSZ of an entire sample. However, as pointed out by Waschkies et al. even with constant solidification velocity, the thickness of the ice crystals does increase slightly over the course of freezing.

Anisotropy of the Interface Kinetics

Even if the temperature gradient within the slurry is perfectly vertical, it is common to see tilting or curvature of the lamellae as they grow through the suspension. To explain this, it is possible to define two distinct growth directions for each ice crystal. There is the direction determined by the temperature gradient, and the one defined by the preferred growth direction crystallographically speaking. These angles are often at odds with one another, and their balance will describe the tilt of the crystal.

The non-overlapping growth directions also help to explain why dendritic textures are often seen in freeze-casts. This texturing is usually found only on the side of each lamella; the direction of the imposed temperature gradient. The ceramic structure left behind shows the negative image of these dendrites. In 2013, Deville et al. made the observation that the periodicity of these dendrites (tip-to-tip distance) actually seems to be related to the primary crystal thickness.

Particle Packing Effects

Up until now, the focus has been mostly on the structure of the ice itself; the particles are almost an afterthought to the templating process but in fact, the particles can and do play a significant role during freeze-casting. It turns out that particle arrangement also changes as a function of the freezing conditions. For example, researchers have shown that freezing velocity has a marked effect on wall roughness. Faster freezing rates produce rougher walls since particles are given insufficient time to rearrange. This could be of use when developing permeable gas transfer membranes where tortuosity and roughness could impede gas flow. It also turns out that z- and r-crystals do not interact with ceramic particles in the same way. The z-crystals pack particles in the x-y plane while r-crystals pack particles primarily in the z-direction. R-crystals actually pack particles more efficiently than z-crystals and because of this, the area fraction of the particle-rich phase (1 - area fraction of ice crystals) changes as the crystal population shifts from a mixture of z- and r-crystals to only z-crystals. Starting from where ice crystals first begin to exclude particles, marking the beginning of the transition zone, we have a majority of r-crystals and a high value for the particle-rich phase fraction. We can assume that because the solidification speed is still rapid that the particles will not be packed efficiently. As the solidification rate slows down, however, the area fraction of the particle-rich phase drops indicating an increase in packing efficiency. At the same time, the competitive growth process is taking place, replacing r-crystals with z-crystals. At a certain point nearing the end of the transition zone, the particle-rich phase fraction rises sharply since z-crystals are less efficient at packing particles than r-crystals. The apex of this curve marks the point where only z-crystals are present (SSZ). During steady-state growth, after the maximum particle-rich phase fraction is reached, the efficiency of packing increases as steady-state is achieved.

In 2011, researchers at Yale University set out to probe the actual spatial packing of particles within the walls. Using small-angle X-ray scattering (SAXS) they characterized the particle size, shape and interparticle spacing of nominally 32 nm silica suspensions that had been freeze-cast at different speeds. Computer simulations indicated that for this system, the particles within the walls should not be touching but rather separated from one another by thin films of ice. Testing, however, revealed that the particles were, in fact, touching and more than that, they attained a packed morphology that cannot be explained by typical equilibrium densification processes.

Morphological Instabilities

In an ideal world, the spatial concentration of particles within the SSZ would remain constant throughout solidification. As it happens, though, the concentration of particles does change during compression, and this process is highly sensitive to solidification speed. At low freezing rates, Brownian motion takes place, allowing particles to move easily away from the solid-liquid interface and maintain a homogeneous suspension. In this situation, the suspension is always warmer than the solidified portion. At fast solidification speeds, approaching VC, the concentration, and concentration gradient at the solid-liquid interface increases because particles cannot redistribute soon enough. When it has built up enough, the freezing point of the suspension is below the temperature gradient in the solution and morphological instabilities can occur. For situations where the particle concentration bleeds into the diffusion layer, both the actual and freezing temperature dip below the equilibrium freezing temperature creating an unstable system. Often, these situations lead to the formation of what are known as ice lenses.

These morphological instabilities can trap particles, preventing full redistribution and resulting in inhomogeneous distribution of solids along the freezing direction as well as discontinuities in the ceramic walls, creating voids larger than intrinsic pores within the walls of the porous ceramic.

Novel Freeze-casting Techniques

Freeze-casting can be applied to numerous materials systems including ceramics, polymers, and metals. As long as there are particles that may be excluded when the solvent changes phase, a templated structure is possible. Using various novel processing techniques, some authors have demonstrated even greater levels of control made available with freeze-casting. Munch et al. showed that it is possible to control the long-range arrangement and orientation of crystals normal to the growth direction by templating the nucleation surface. This technique works by providing lower energy nucleation sites to control the initial crystal growth and arrangement. The orientation of ice crystals can also be affected by applied electromagnetic fields as was demonstrated in 2010 by Tang et al. Using specialized setups, researchers have been able to create radially aligned freeze-casts tailored for filtration or gas separation applications. Inspired by Nature, scientists have also been able to use coordinating chemicals and cryopreserved to create remarkably distinctive microstructural architectures.

COMPACTION OF CERAMIC POWDERS

Compaction of ceramic powders is a specific forming technique for ceramics. It is a process in which ceramic granular materials are made cohesive through mechanical densification, involving (hot pressing) or not (cold forming) temperature exposition. The process permits an efficient production of parts ranging widely in size and shape to close tolerances with low drying shrinkage. Traditional (for instance: Ceramic tiles, porcelain products) and structural (for instance: Chip carriers, spark plugs, cutting tools) ceramics are produced. Cold compaction of ceramic powders ends up with the realization of the so-called green piece, which is later subject to sintering.

Ceramic industry is broadly developed in the world, so much so that only in Europe the current investment is estimated at € 26 billion. Advanced ceramics are crucial for new technologies, addressed particularly to thermo-mechanical and bio-medical applications, while traditional ceramics have a worldwide market and have been suggested as materials minimizing the impact on the environment (when compared to other finishing materials).

Production Process of Ceramics

Up-to-date ceramic technology involves invention and design of new components and optimization of production process of complex structures. Ceramics can be formed by a variety of different methods which can be divided into three main groups, depending on whether the starting materials involve a gas, a liquid or a solid. Examples of methods involving gases are: chemical vapour deposition, directed metal oxidation and reaction bonding, whereas examples of methods involving liquids are: sol-gel process and polymer pyrolysis. Methods involving solids, and especially powder methods, dominate ceramic forming and are extensively used in the industry.

The practical realization of ceramic products by powder methods requires the following steps: ceramic powder production, powder treatment, handling and processing, cold forming, sintering and evaluation of the performance of the final product. Since this process permits an efficient production of parts ranging widely in size and shape to close tolerances, there is an evident interest from industry. For instance, metallurgical, pharmaceutical and forming of traditional and advanced structural ceramics represent common applications.

Mechanics of Forming of Ceramic Powders

It is a well-established fact that the performance of a ceramic component critically depends on the manufacturing process. In particular, initial powder characteristics and processing, including cold forming and sintering, have a strong impact on the mechanical properties of the components as they may generate a defect population (microcracks, density gradients, pores, agglomerates) within the green and sintered compounds. In particular, the mechanical characteristics of the solid obtained after cold forming (the so-called 'green body') strongly affect the subsequent sintering process and thus the mechanical properties of the final piece.

A piece (formed with M KMS-96 alumina powder)
has been broken after mold ejection.

Many technical, still unresolved difficulties arise in the forming process of ceramic materials. In fact, if on one hand the compact should result intact after ejection, should be handleable without failure and essentially free of macro defects, on the other hand, defects of various nature are always present in the green bodies, negatively affecting local shrinkage during sintering.

Defects can be caused by densification process, which may involve highly inhomogeneous strain fields, or by mold ejection. Currently, there is a high production reject, due to the fact that manufacturing technologies are mainly based on empirically engineered processes, rather than on rational and scientific methodologies.

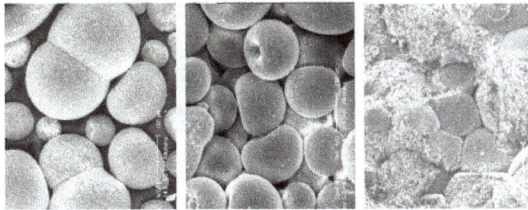

Micrographs of a M KMS-96 alumina powder.

The loose state is shown on the left, while granule arrangements corresponding to Phases I and II of the compaction process are shown on the centre and on the right. Note the plastic deformation of grains visible on the right.

The industrial technologies involved in the production of ceramics, with particular reference to tile and sanitaryware products, are directly connected to a huge amount of waste of material and energy. Consequently, the set-up of manufacturing processes is very costly and time consuming and not yet optimal in terms of quality of the final piece.

There is therefore a strong interest from the ceramic industry in the availability of tools capable of modelling and simulating: i) the powder compaction process and ii) the criticality of defects possibly present in the final piece after sintering. Recently, an EU IAPP research project has been financed with the aim of enhance mechanical modelling of ceramic forming in view of industrial applications.

During cold powder compaction, a granular material is made cohesive through mechanical densification, a process which modelling requires the description of the transition from a granular to a dense and even a fully dense state.

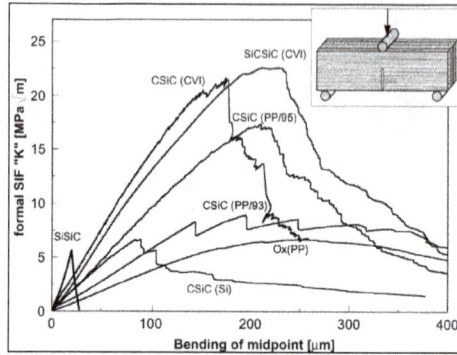

The hardening process during hydrostatic powder compaction
described with the Bigoni & Piccolroaz yield surface.

Since granular materials are characterized by mechanical properties almost completely different from those typical of dense solids, the mechanical modelling must describe a transition between two distinctly different states of a material. This is a scientific challenge addressed by Piccolroaz et al. in terms of plasticity theory.

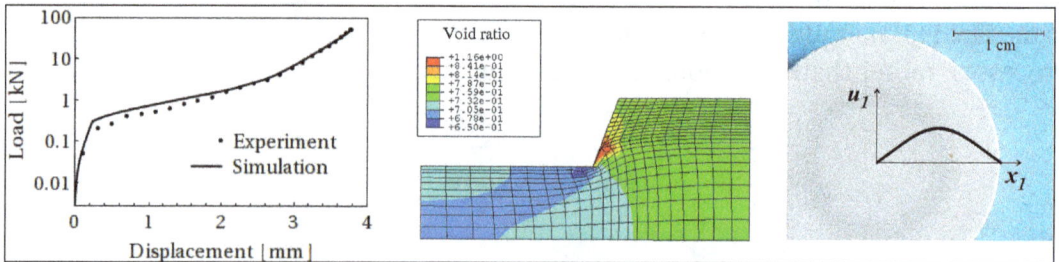

A mechanical model of ceramic forming correctly predicts: (left) the load/displacement curve during cold pressing, (centre) the density (void ratio) map within a formed piece and (right) the dark annular region evidenced on the bottom of a formed piece.

The mechanical model developed by Piccoloraz et al. permits the description of the forming process. The INTERCER2 research project is aimed to develop novel constitutive descriptions for ceramic powders and more robust implementation in a numerical code.

CERAMIC MOLD CASTING

Ceramic mold casting, also known ambiguously as ceramic molding, is a group of metal casting processes that use ceramics as the mold material. It is a combination of plaster mold casting and investment casting. There are two types of ceramic mold casting: The *Shaw process* and the *Unicast process*.

These casting processes are commonly used to make tooling, especially drop forging dies, but also injection molding dies, die casting dies, glass molds, stamping dies, and extrusion dies.

Shaw Process

The *Shaw process*, also known as the *Osborn-Shaw process*, uses a mixture of refractory aggregate, hydrolyzed ethyl silicate, alcohol, and a gelling agent to create a mold. This slurry mixture is poured into a slightly tapered flask and a reusable pattern (*i.e.* the item used to create the shape of the mold) is used. The slurry hardens almost immediately to a rubbery state (the consistency of vulcanized rubber). The flask and pattern is then removed. Then a torch is used to ignite the mold, which causes most of the volatiles to burn-off and the formation of ceramic microcrazes (microscopic cracks). These cracks are important, because they allow gases to escape while preventing the metal from flowing through; they also ease thermal expansion and contraction during solidification and shrinkage. After the burn-off, the mold is baked at 1,800 °F (980 °C) to remove any remaining volatiles. Prior to pouring metal, the mold is pre-warmed to control shrinkage.

Unicast Process

The *Unicast process* is very similar to the Shaw process, except it does not require the mold to be ignited and then be cured in a furnace. Instead, the mold is partially cured so the pattern can be removed and it is then completely cured by firing it at approximately 1,900 °F (1,040 °C). If a metal with a low melting point is cast then the firing can be skipped, because the mold has enough strength in the "green state" (un-fired).

Characteristics

The main advantages of ceramic molds are: a reusable pattern (the item used to create the shape of the mold), excellent surface finish, close dimensional tolerances, thin cross-sections, and intricate shapes can be cast. For undercuts and other difficult to cast features, part of the pattern can be made from wax in conjunction with a standard pattern; essentially using investment and ceramic mold casting techniques together. The main disadvantages are: it is only cost effective for small- to medium-sized production runs and the ceramic is not reusable. Ferrous and high-temperature non-ferrous are most commonly cast with these processes; other materials cast include: aluminum, copper, magnesium, titanium, and zinc alloys.

Weight limits are 100 grams to several thousand kilograms (3.5 oz to several tons). Cross-sections as thin as 1.3 mm (0.051 in) are possible, with no upper limit. Typical tolerances are 0.1 mm for the first 25 mm (0.005 in for the first inch) and 0.003 mm per additional mm (0.003 in per each additional in). A draft of 1° is typically required. The typical surface finish is 2–4 um (75–150 uin) RMS.

CERAMIC MATRIX COMPOSITE

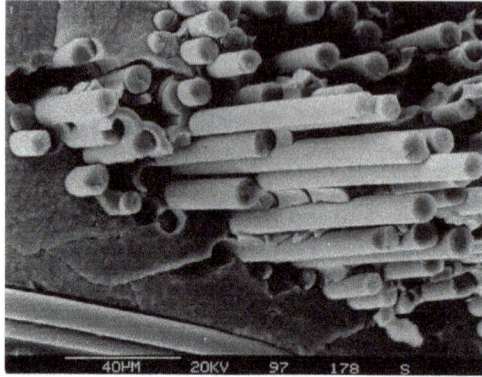

Fracture surface of a fiber-reinforced ceramic composed of SiC fibers and SiC matrix. The fiber pull-out mechanism shown is the key to CMC properties.

CMC shaft sleeves.

Ceramic matrix composites (CMCs) are a subgroup of composite materials as well as a subgroup of ceramics. They consist of ceramic fibres embedded in a ceramic matrix. The matrix and fibres can consist of any ceramic material, whereby carbon and carbon fibres can also be considered a ceramic material.

The motivation to develop CMCs was to overcome the problems associated with the conventional technical ceramics like alumina, silicon carbide, aluminium nitride, silicon nitride or zirconia – they fracture easily under mechanical or thermo-mechanical loads because of cracks initiated by small defects or scratches. The crack resistance is – like in glass – very low. To increase the crack resistance or fracture toughness, particles (so-called monocrystalline *whiskers* or *platelets*) were embedded into the matrix. However, the improvement was limited, and the products have found application only in some ceramic cutting tools. So far only the integration of long multistrand fibres has drastically increased the crack resistance, elongation and thermal shock resistance, and resulted in several new applications. The reinforcements used in ceramic matrix composites (CMC) serve to enhance the fracture toughness of the combined material system while still taking advantage of the inherent high strength and Young's modulus of the ceramic matrix. The most common reinforcement embodiment is a continuous-length ceramic fiber, with an elastic modulus that is typically somewhat higher than the matrix. The functional role of this fiber is (1) to increase the CMC stress for progress of micro-cracks through the matrix, thereby increasing the energy expended during crack propagation; and then (2) when thru-thickness cracks begin to form across the CMC at a higher stress (proportional limit stress, PLS), to bridge these cracks without fracturing, thereby providing the CMC with a high ultimate tensile strength (UTS). In this way, ceramic fiber reinforcements not only increase the composite structure's initial resistance to crack propagation, but also allow the CMC to avoid abrupt brittle failure that is characteristic of monolithic ceramics. This behavior is distinct from the behavior of ceramic fibers in polymer matrix composites (PMC) and metal matrix composites (MMC), where the fibers typically fracture prior to the matrix due to the higher failure strain capabilities of these matrices.

Carbon (C), special silicon carbide (SiC), alumina (Al_2O_3) and mullite ($Al_2O_3–SiO_2$) fibres are most commonly used for CMCs. The matrix materials are usually the same, that is C, SiC, alumina and mullite. Recently Ultra-high-temperature ceramics (UHTCs) were investigated as ceramic matrix in a new class of CMC so-called Ultra-high Temperature Ceramic Matrix Composites (UHTCMC) or Ultra-high Temperature Ceramic Composites (UHTCC).

Generally, CMC names include a combination of *type of fibre/type of matrix*. For example, *C/C* stands for carbon-fibre-reinforced carbon (carbon/carbon), or *C/SiC* for carbon-fibre-reinforced silicon carbide. Sometimes the manufacturing process is included, and a C/SiC composite manufactured with the liquid polymer infiltration (LPI) process is abbreviated as *LPI-C/SiC*.

The important commercially available CMCs are C/C, C/SiC, SiC/SiC and Al_2O_3/Al_2O_3. They differ from conventional ceramics in the following properties:

1. Elongation to rupture up to 1%.

2. Strongly increased fracture toughness.

3. Extreme thermal shock resistance.

4. Improved dynamical load capability.

5. Anisotropic properties following the orientation of fibers.

Manufacture

The manufacturing processes usually consist of the following three steps:

1. Lay-up and fixation of the fibres, shaped as the desired component.

2. Infiltration of the matrix material.

3.   Final machining and, if required, further treatments like coating or impregnation of the intrinsic porosity.

The first and the last step are almost the same for all CMCs: In step one, the fibres, often named rovings, are arranged and fixed using techniques used in fibre-reinforced plastic materials, such as lay-up of fabrics, filament winding, braiding and knotting. The result of this procedure is called *fibre-preform* or simply *preform*.

For the second step, five different procedures are used to fill the ceramic matrix in between the fibres of the preform:

1.   Deposition out of a gas mixture.

2.   Pyrolysis of a pre-ceramic polymer.

3.   Chemical reaction of elements.

4.   Sintering at a relatively low temperature in the range 1000–1200 °C.

5.   Electrophoretic deposition of a ceramic powder.

Procedures one, two and three find applications with non-oxide CMCs, whereas the fourth one is used for oxide CMCs; combinations of these procedures are also practised. The fifth procedure is not yet established in industrial processes. All procedures have sub-variations, which differ in technical details. All procedures yield a porous material.

The third and final step of machining – grinding, drilling, lapping or milling – has to be done with diamond tools. CMCs can also be processed with a water jet, laser, or ultrasonic machining.

Ceramic Fibres

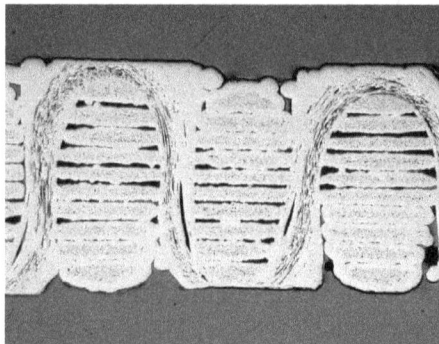

Micrograph of a SiC/SiC ceramic composite with
a woven three-dimensional fibre structure.

Ceramic fibres in CMCs can have a polycrystalline structure, as in conventional ceramics. They can also be amorphous or have inhomogeneous chemical composition, which develops upon pyrolysis of organic precursors. The high process temperatures required for making CMCs preclude the use of organic, metallic or glass fibres. Only fibres stable at temperatures above 1000 °C can be used, such as fibres of alumina, mullite, SiC, zirconia or carbon. Amorphous SiC fibres have an elongation capability above 2% – much larger than in conventional ceramic materials (0.05 to 0.10%). The reason for this property of SiC fibres is that most of them contain additional elements like oxygen, titanium and/or aluminium yielding a tensile strength above 3 GPa. These enhanced elastic properties are required for various three-dimensional fibre arrangements in textile fabrication, where a small bending radius is essential.

Manufacturing Procedures

Matrix Deposition from a Gas Phase

Chemical vapour deposition (CVD) is well suited for this purpose. In the presence of a fibre preform, CVD takes place in between the fibres and their individual filaments and therefore is called chemical vapour infiltration (CVI). One example is the manufacture of C/C composites: A C-fibre preform is exposed to a mixture of argon and a hydrocarbon gas (methane, propane, etc.) at a pressure of around or below 100 kPa and a temperature above 1000 °C. The gas decomposes depositing carbon on and between the fibers. Another example is the deposition of silicon carbide, which is usually conducted from a mixture of hydrogen and methyl-trichlorosilane (MTS, CH_3SiCl_3; it is also common in silicone production). Under defined condition this gas mixture deposits fine and crystalline silicon carbide on the hot surface within the preform. This CVI procedure leaves a body with a porosity of about 10–15%, as access of reactants to the interior of the preform is increasingly blocked by deposition on the exterior.

Matrix Forming via Pyrolysis of C- and Si-containing Polymers

Hydrocarbon polymers shrink during pyrolysis, and upon outgassing form carbon with an amorphous, glass-like structure, which by additional heat treatment can be changed to a more graphite-like structure. Other special polymers, known as preceramic polymers where some carbon atoms are replaced by silicon atoms, the so-called polycarbosilanes, yield amorphous silicon carbide of more or less stoichiometric composition. A large variety of such silicon carbide, silicon oxycarbide, silicon carbonitride and silicon oxynitride precursors already exist and more preceramic polymers for the fabrication of polymer derived ceramics are being developed. To manufacture a CMC material, the fibre preform is infiltrated with the chosen polymer. Subsequent curing and pyrolysis yield a highly porous matrix, which is undesirable for most applications. Further cycles of polymer infiltration and pyrolysis are performed until the final and desired quality is achieved. Usually five to eight cycles are necessary. The process is called *liquid polymer infiltration* (LPI), or *polymer infiltration and pyrolysis* (PIP). Here also a porosity of about 15% is common due to the shrinking of the polymer. The porosity is reduced after every cycle.

Matrix Forming via Chemical Reaction

With this method, one material located between the fibres reacts with a second material to form the ceramic matrix. Some conventional ceramics are also manufactured by chemical reactions. For example, reaction-bonded silicon nitride (RBSN) is produced through the reaction of silicon powder with nitrogen, and porous carbon reacts with silicon to form reaction bonded silicon carbide, a silicon carbide which contains inclusions of a silicon phase. An example of CMC manufacture, which was introduced for the production of ceramic brake discs, is the reaction of silicon with a porous preform of C/C. The process temperature is above 1414 °C, that is above the melting point of silicon, and the process conditions are controlled such that the carbon fibres of the C/C-preform almost completely retain their mechanical properties. This process is called *liquid silicon infiltration* (LSI). Sometimes, and because of its starting point with C/C, the material is abbreviated as *C/C-SiC*. The material produced in this process has a very low porosity of about 3%.

Matrix Forming via Sintering

This process is used to manufacture oxide fibre/oxide matrix CMC materials. Since most ceramic fibres cannot withstand the normal sintering temperatures of above 1600 °C, special precursor liquids are used to infiltrate the preform of oxide fibres. These precursors allow sintering, that is ceramic-forming processes, at temperatures of 1000–1200 °C. They are, for example, based on mixtures of alumina powder with the liquids tetra-ethyl-orthosilicate (as Si donor) and aluminium-butylate (as Al donor), which yield a mullite matrix. Other techniques, such as sol-gel chemistry, are also used. CMCs obtained with this process usually have a high porosity of about 20%.

Matrix Formed via Electrophoresis

In the electrophoretic process, electrically charged particles dispersed in a special liquid are transported through an electric field into the preform, which has the opposite electrical charge polarity. This process is under development, and is not yet used industrially. Some remaining porosity must be expected here, too.

Properties

Scheme of crack bridges at the crack tip of ceramic composites.

Mechanical Properties

Basic Mechanism of Mechanical Properties

The high fracture toughness or crack resistance mentioned above is a result of the following mechanism: under load the ceramic matrix cracks, like any ceramic material, at an elongation of about 0.05%. In CMCs the embedded fibres bridge these cracks. This mechanism works only when the matrix can slide along the fibres, which means that there must be a weak bond between the fibres and matrix. A strong bond would require a very high elongation capability of the fibre bridging the crack, and would result in a brittle fracture, as with conventional ceramics. The production of CMC material with high crack resistance requires a step to weaken this bond between the fibres and matrix. This is achieved by depositing a thin layer of pyrolytic carbon or boron nitride on the fibres, which weakens the bond at the fibre/matrix interface (sometimes "interface"), leading to the fibre pull-out at crack surfaces, In oxide-CMCs, the high porosity of the matrix is sufficient to establish the weak bond.

Properties under Tensile and Bending Loads, Crack Resistance

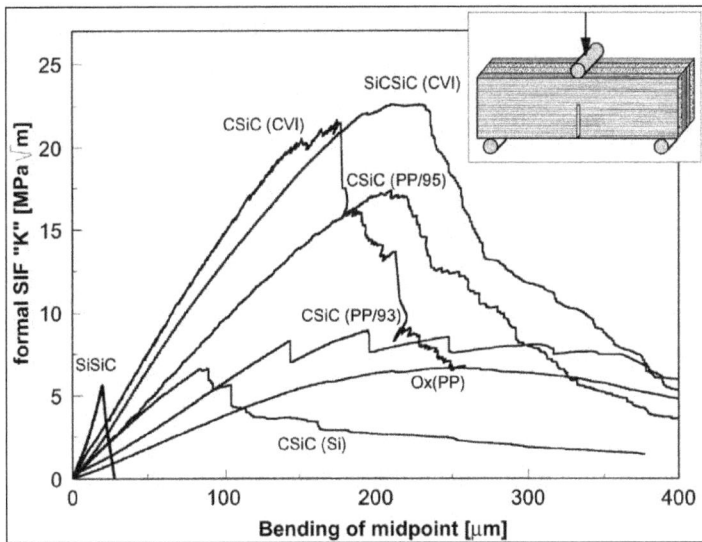

Curves of toughness measurements of various ceramic composites and SiSiC.

The influence and quality of the fibre interface can be evaluated through mechanical properties. Measurements of the crack resistance were performed with notched specimens in so-called single-edge-notch-bend (SENB) tests. In fracture mechanics, the measured data (force, geometry and crack surface) are normalized to yield the so-called stress intensity factor (SIF), K_{Ic}. Because of the complex crack surface the real crack surface area can not be determined for CMC materials. The measurements therefore use the initial notch as the crack surface, yielding the *formal SIF* shown in the figure. This requires identical geometry for comparing different samples. The area under these curves thus gives a relative indication of the energy required to drive the crack tip through the sample (force times path length gives energy). The maxima indicate the load level necessary to propagate the crack through the sample. Compared to the sample of conventional SiSiC ceramic, two observations can be made:

- All tested CMC materials need up to several orders of magnitude more energy to propagate the crack through the material.

- The force required for crack propagation varies between different types of CMCs.

Type of material	Al_2O_3/Al_2O_3	Al_2O_3	CVI-C/SiC	LPI-C/SiC	LSI-C/SiC	SiSiC
Porosity (%)	35	<1	12	12	3	<1
Density (g/cm³)	2.1	3.9	2.1	1.9	1.9	3.1
Tensile strength (MPa)	65	250	310	250	190	200
Elongation (%)	0.12	0.1	0.75	0.5	0.35	0.05
Young's modulus (GPa)	50	400	95	65	60	395
Flexural strength (MPa)	80	450	475	500	300	400

In the table, CVI, LPI, and LSI denote the manufacturing process of the C/SiC-material. Data of the oxide CMC and SiSiC are taken from manufacturer data sheets. Tensile strength of SiSiC and Al_2O_3 were calculated from measurements of elongation to fracture and Young's modulus, since generally only bending strength data are available for those ceramics. Averaged values are given in the table, and significant differences, even within one manufacturing route, are possible.

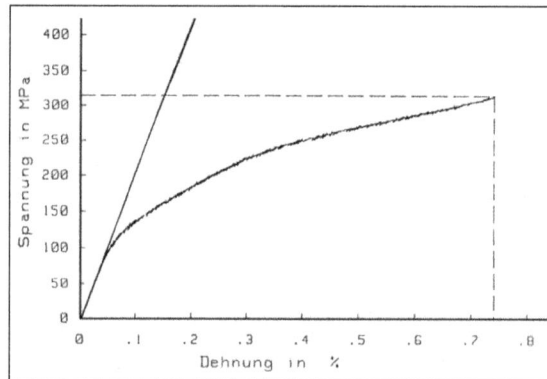

Stress-strain curve of a tensile test for CVI-SiC/SiC.

Tensile tests of CMCs usually show nonlinear stress-strain curves, which look as if the material deforms plastically. It is called *quasi-plastic*, because the effect is caused by the microcracks, which are formed and bridged with increasing load. Since the Young's modulus of the load-carrying fibres is generally lower than that of the matrix, the slope of the curve decreases with increasing load.

Curves from bending tests look similar to those of the crack resistance measurements shown above.

The following features are essential in evaluating bending and tensile data of CMCs:

- CMC materials with a low matrix content (down to zero) have a high tensile strength (close to the tensile strength of the fibre), but low bending strength.

- CMC materials with a low fibre content (down to zero) have a high bending strength (close to the strength of the monolithic ceramic), but no elongation beyond 0.05% under tensile load.

The primary quality criterion for CMCs is the crack resistance behaviour or fracture toughness.

Other Mechanical Properties

In many CMC components the fibres are arranged as 2-dimensional (2D) stacked plain or satin weave fabrics. Thus the resulting material is anisotropic or, more specifically, orthotropic. A crack between the layers is not bridged by fibres. Therefore, the interlaminar shear strength (ILS) and the strength perpendicular to the 2D fiber orientation are low for these materials. Delamination can occur easily under certain mechanical loads. Three-dimensional fibre structures can improve this situation.

Material	CVI-C/SiC	LPI-C/SiC	LSI-C/SiC	CVI-SiC/SiC
Interlaminar shear strength (MPa)	45	30	33	50
Tensile strength vertical to fabric plane (MPa)	6	4	–	7
Compressive strength vertical to fabric plane (MPa)	500	450	–	500

The compressive strengths shown in the table are lower than those of conventional ceramics, where values above 2000 MPa are common; this is a result of porosity.

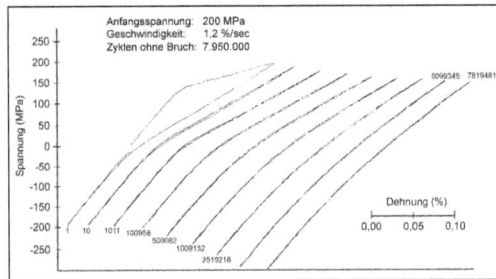

Strain controlled LCF-test for a CVI-SiC/SiC-specimen.

The composite structure allows high dynamical loads. In the so-called low-cycle-fatigue (LCF) or high-cycle-fatigue (HCF) tests the material experiences cyclic loads under tensile and compressive (LCF) or only tensile (HCF) load. The higher the initial stress the shorter the lifetime and the smaller the number of cycles to rupture. With an initial load of 80% of the strength, a SiC/SiC sample survived about 8 million cycles.

The Poisson's ratio shows an anomaly when measured perpendicular to the plane of the fabric, because interlaminar cracks increase the sample thickness.

Thermal and Electrical Properties

The thermal and electrical properties of the composite are a result of its constituents, namely fibres, matrix and pores as well as their composition. The orientation of the fibres yields anisotropic data. Oxide CMCs are very good electrical insulators, and because of their high porosity their thermal insulation is much better than that of conventional oxide ceramics.

The use of carbon fibres increases the electrical conductivity, provided the fibres contact each other and the voltage source. Silicon carbide matrix is a good thermal conductor. Electrically, it is a semiconductor, and its resistance therefore decreases with increasing temperature. Compared to (poly)crystalline SiC, the amorphous SiC fibres are relatively poor conductors of heat and electricity.

Material	CVI-C/SiC	LPI-C/SiC	LSI-C/SiC	CVI-SiC/SiC	SiSiC
Thermal conductivity (p) [W/(m·K)]	15	11	21	18	>100
Thermal conductivity (v) [W/(m·K)]	7	5	15	10	>100
Linear expansion (p) [10^{-6}·1/K]	1.3	1.2	0	2.3	4
Linear expansion (v) [10^{-6}·1/K]	3	4	3	3	4
Electrical resistivity (p) [Ω·cm]	–	–	–	–	50
Electrical resistivity (v) [Ω·cm]	0.4	–	–	5	50

Table: (p) and (v) refer to data parallel and vertical to fibre orientation of the 2D-fiber structure, respectively. LSI material has the highest thermal conductivity because of its low porosity – an advantage when using it for brake discs. These data are subject to scatter depending on details of the manufacturing processes.

Conventional ceramics are very sensitive to thermal stress because of their high Young's modulus and low elongation capability. Temperature differences and low thermal conductivity create locally different elongations, which together with the high Young's modulus generate high stress. This results in cracks, rupture and brittle failure. In CMCs, the fibres bridge the cracks, and the components show no macroscopic damage, even if the matrix has cracked locally. The application of CMCs in brake disks demonstrates the effectiveness of ceramic composite materials under extreme thermal shock conditions.

Corrosion Properties

Data on the corrosion behaviour of CMCs are scarce except for oxidation at temperatures above 1000 °C. These properties are determined by the constituents, namely the fibres and matrix. Ceramic materials in general are very stable to corrosion. The broad spectrum of manufacturing techniques with different sintering additives, mixtures, glass phases and porosities are crucial for the results of corrosion tests. Less impurities and exact stoichiometry lead to less corrosion. Amorphous structures and non-ceramic chemicals frequently used as sintering aids are starting points of corrosive attack.

Alumina

Pure alumina shows excellent corrosion resistivity against most chemicals. Amorphous glass and silica phases at the grain boundaries determine the speed of corrosion in concentrated acids and bases and result in creep at high temperatures. These characteristics limit the use of alumina. For molten metals, alumina is used only with gold and platinum.

Alumina Fibres

These fibres demonstrate corrosion properties similar to alumina, but commercially available fibres are not very pure and therefore less resistant. Because of creep at temperatures above 1000 °C, there are only few applications for oxide CMCs.

Carbon

The most significant corrosion of carbon occurs in presence of oxygen above about 500 °C. It burns to form carbon dioxide and/or carbon monoxide. It also oxidises in strong oxidizing agents like concentrated nitric acid. In molten metals it dissolves and forms metal carbides. Carbon fibres do not differ from carbon in their corrosion behaviour.

Silicon Carbide

Pure silicon carbide is one of the most corrosion-resistant materials. Only strong bases, oxygen above about 800 °C, and molten metals react with it to form carbides and silicides. The reaction with oxygen forms SiO_2 and CO_2, whereby a surface layer of SiO_2 slows down subsequent oxidation (*passive oxidation*). Temperatures above about 1600 °C and a low partial pressure of oxygen result in so-called *active oxidation*, in which CO, CO_2 and gaseous SiO are formed causing rapid loss of SiC. If the SiC matrix is produced other than by CVI, corrosion-resistance is not as good. This is a consequence of porosity in the amorphous LPI, and residual silicon in the LSI-matrix.

Silicon Carbide Fibres

Silicon carbide fibres are produced via pyrolysis of organic polymers, and therefore their corrosion properties are similar to those of the silicon carbide found in LPI-matrices. These fibres are thus more sensitive to bases and oxidizing media than pure silicon carbide.

Applications

CMC materials overcome the major disadvantages of conventional technical ceramics, namely brittle failure and low fracture toughness, and limited thermal shock resistance.

Therefore, their applications are in fields requiring reliability at high-temperatures (beyond the capability of metals) and resistance to corrosion and wear. These include:

- Heat shield systems for space vehicles, which are needed during the re-entry phase, where high temperatures, thermal shock conditions and heavy vibration loads take place.

- Components for high-temperature gas turbines such as combustion chambers, stator vanes and turbine blades.

- Components for burners, flame holders, and hot gas ducts, where the use of oxide CMCs has found its way.

- Brake disks and brake system components, which experience extreme thermal shock (greater than throwing a glowing part of any material into water).

- Components for slide bearings under heavy loads requiring high corrosion and wear resistance.

In addition to the foregoing, CMCs can be used in applications, which employ conventional ceramics or in which metal components have limited lifetimes due to corrosion or high temperatures.

Developments for Applications in Space

During the re-entry phase of space vehicles, the heat shield system is exposed to temperatures above 1500 °C for a few minutes. Only ceramic materials are able to survive such conditions without significant damage, and among ceramics only CMCs can adequately handle thermal shocks. The development of CMC-based heat shield systems promises the following advantages:

- Reduced weight.

- Higher load carrying capacity of the system.

- Reusability for several re-entries.

- Better steering during the re-entry phase with CMC flap systems.

NASA-space vehicle X-38 during a test flight.

In these applications the high temperatures preclude the use of oxide fibre CMCs, because under the expected loads the creep would be too high. Amorphous silicon carbide fibres lose their strength due to re-crystallization at temperatures above 1250 °C. Therefore, carbon fibres in a silicon carbide matrix (C/SiC) are used in development programs for these applications. The European program HERMES of ESA, started in the 1980s and for financial reasons abandoned in 1992, has produced first results. Several follow-up programs focused on the development, manufacture, and qualification of nose cap, leading edges and steering flaps for the NASA space vehicle X-38.

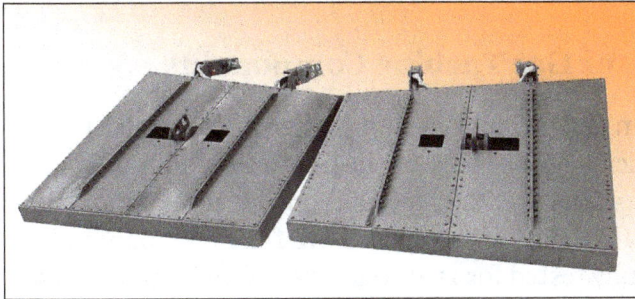

Pair of steering flaps for the NASA-space vehicle X-38. Size: 1.5×1.5×0.15 m, mass: 68 kg each, various components are mounted using more than 400 CVI-C/SiC screws and nuts.

This development program has qualified the use of C/SiC bolts and nuts, and the bearing system of the flaps. The latter were ground-tested at the DLR in Stuttgart, Germany, under expected conditions of the re-entry phase: 1600 °C, 4 tonnes load, oxygen partial pressure similar to re-entry conditions, and simultaneous bearing movements of four cycles per second. A total of five re-entry phases was simulated. Furthermore, oxidation protection systems were developed and qualified to prevent burnout of the carbon fibres. After mounting of the flaps, mechanical ground tests were performed successfully by NASA in Houston, Texas, US. The next test – a real re-entry of the unmanned vehicle X-38 – was cancelled for financial reasons. One of the space shuttles would have brought the vehicle into orbit, from where it would have returned to the Earth.

These qualifications were promising for only this application. The high-temperature load lasts only around 20 minutes per re-entry, and for reusability, only about 30 cycles would be sufficient. For industrial applications in hot gas environment, though, several hundred cycles of thermal loads and up to many thousands hours of lifetime are required.

The Intermediate Experimental Vehicle (IXV), a project initiated by ESA in 2009, is Europe's first lifting body reentry vehicle. Developed by Thales Alenia Space, the IXV is scheduled to make its first flight in 2014 on the fourth Vega mission (VV04) over the Gulf of Guinea. More than 40 European companies contributed to its construction. The thermal protection system for the underside of the vehicle, comprising the nose, leading edges and lower surface of the wing, were designed and made by Herakles using a ceramic matrix composite (CMC), carbon/silicon-carbide (C/SiC). These components will function as the vehicle's heat shield during its atmospheric reentry.

The European Commission funded a research project, C3HARME, under the NMP-19-2015 call of Framework Programmes for Research and Technological Development (H2020) in 2016 for the design, development, production and testing of a new class of Ultra-high-temperature ceramic matrix composites (UHTCMC) reinforced with silicon carbide fibers and Carbon fibers suitable for applications in severe aerospace environments such as propulsion and Thermal protection systems (TPSs).

Developments for Gas Turbine Components

The use of CMCs in gas turbines permit higher turbine inlet temperatures, which improves turbine efficiency. Because of the complex shape of stator vanes and turbine blades, the development was first focused on the combustion chamber. In the US, a combustor made of SiC/SiC with a special SiC fiber of enhanced high-temperature stability was successfully tested for 15,000 hours. SiC oxidation was substantially reduced by the use of an oxidation protection coating consisting of several layers of oxides.

The engine collaboration between General Electric and Rolls-Royce studied the use of CMC stator vanes in the hot section of the F136 turbofan engine, an engine which failed to beat the Pratt and Whitney F-135 for use in the Joint Strike Fighter. The engine joint venture, CFM International is using CMCs to manufacture the high temperature turbine shrouds. General Electric is using CMCs in combustor liners, nozzles, and the high temperature turbine shroud for its upcoming GE9X engine. CMC parts are also being studied for stationary applications in both the cold and hot sections of the engines, since stresses imposed on rotating parts would require further development effort. Generally, development continues of CMCs for use in turbines to reduce technical issues and cost reduction.

After $1.5 billion in investment and 20 years of research and development, by 2020 GE Aviation aims to produce per year up to 20 t (44,000 lb) of CMC prepreg and 10 t of silicon carbide fiber. Chemical vapor deposition can apply coatings on a laid-able fiber tape in large quantities and Ge managed to infiltrate and cast parts with very high silicon densities, higher than 90% for cyclic fatigue environments, thanks to thermal processing.

Application of Oxide CMC in Burner and Hot Gas Ducts

Oxygen-containing gas at temperatures above 1000 °C is rather corrosive for metal and silicon carbide components. Such components, which are not exposed to high mechanical stress, can be made of oxide CMCs, which can withstand temperatures up to 1200 °C. The gallery below shows the flame holder of a crisp bread bakery as tested after for 15,000 hours, which subsequently operated for a total of more than 20,000 hours.

| Oxide CMC flame holder. | Ventilator for hot gases. | Lifting gate, oxide CMC. | Lifting gate in the field. |

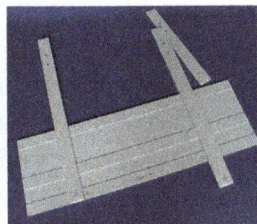

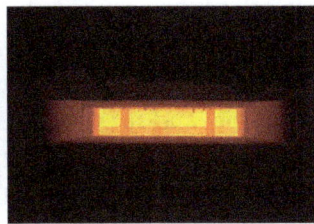

Flaps and ventilators circulating hot, oxygen-containing gases can be fabricated in the same shape as their metal equivalents. The lifetime for these oxide CMC components is several times longer than for metals, which often deform. A further example is an oxide CMC lifting gate for a sintering furnace, which has survived more than 260,000 opening cycles.

Application in Brake Disk

Carbon/carbon (C/C) materials have found their way into the disk brakes of racing cars and aeroplanes, and C/SiC brake disks manufactured by the LSI process were qualified and are commercially available for luxury vehicles. The advantages of these C/SiC disks are:

- Very little wear, resulting in lifetime use for a car with a normal driving load of 300,000 km, is forecast by manufacturers.

- No fading is experienced, even under high load.

- No surface humidity effect on the friction coefficient shows up, as in C/C brake disks.

- The corrosion resistance, for example to the road salt, is much better than for metal disks.

- The disk mass is only 40% of a metal disk. This translates into less unsprung and rotating mass.

The weight reduction improves shock absorber response, road-holding comfort, agility, fuel economy, and thus driving comfort.

The SiC-matrix of LSI has a very low porosity, which protects the carbon fibres quite well. Brake disks do not experience temperatures above 500 °C for more than a few hours in their lifetime. Oxidation is therefore not a problem in this application. The reduction of manufacturing costs will decide the success of this application for middle-class cars.

Application in Slide Bearings

Conventional SiC, or sometimes the less expensive SiC, have been used successfully for more than 25 years in slide or journal bearings of pumps. The pumped liquid itself provides the lubricant for the bearing. Very good corrosion resistance against practically all kinds of media, and very low wear and low friction coefficients are the basis of this success. These bearings consist of a static bearing, shrink-fitted in its metallic environment, and a rotating shaft sleeve, mounted on the shaft. Under compressive stress the ceramic static bearing has a low risk of failure, but a SiC shaft sleeve does not have this situation and must therefore have a large wall thickness and/or be specially designed. In large pumps with shafts 100–350 mm in diameter, the risk of failure is higher due to the changing requirements on the pump performance – for example, load changes during operation. The introduction of SiC/SiC as a shaft sleeve material has proven to be very successful. Test rig experiments showed an almost triple specific load capability of the bearing system with a shaft sleeve made of SiC/SiC, sintered SiC as static bearing, and water at 80 °C as lubricant. The specific load capacity of a bearing is usually given in W/mm^2 and calculated as a product of the load (MPa), surface speed of the bearing (m/s) and friction coefficient; it is equal to the power loss of the bearing system due to friction.

Components for a ceramic slide bearing; the picture shows a sintered SiC-bearing for a hydrostatic slide bearing and a CVI-SiC/SiC-shaft sleeve shrink-fitted on metal, a system tested with liquid oxygen as lubricant.

In boiler feedwater pumps of power stations, which pump several thousand cubic meters of hot water to a level of 2000 m, and in tubular casing pumps for water works or sea water desalination plants (pumping up to 40,000 m^3 to a level of around 20 m) this slide bearing concept, namely SiC/SiC shaft sleeve and SiC bearing, has been used since 1994.

This bearing system has been tested in pumps for liquid oxygen, for example in oxygen turbopumps for thrust engines of space rockets, with the following results. SiC and SiC/SiC are compatible with liquid oxygen. In an auto-ignition test according to the French standard NF 28-763, no auto-ignition was observed with powdered SiC/SiC in 20 bar pure oxygen at temperatures up to 525 °C. Tests have shown that the friction coefficient is half, and wear one fiftieth of standard metals used in this environment. A hydrostatic bearing system has survived several hours at a speed up to 10,000 revolutions per minute, various loads, and 50 cycles of start/stop transients without any significant traces of wear.

Other Applications and Developments

- Thrust control flaps for military jet engines.

- Components for fusion and fission reactors.

- Friction systems for various applications.

- Nuclear applications.

- Heat treatment, high temperature, soldering fixtures.

CERAMIC SHELL MOLDING

Ceramic shell molding is the most rapidly used technique for mold and core making. Also known a croning process, this casting technique. Cermaic mold shell molding was invented and patented by J.Croning durign World War II. Also know as the process, shell molding technique is used for making thin sections and for acquiring surface finish and dimensional accuracy.

Process: In the first stage of ceramic shell molding, a metal pattern is made which is resistant to high temperature and can withstand abrasion due to contact with sand. The sand and resin mixture for the shell mold is brought in contact with the pattern. The mold is placed in an oven where the resin is cured. This process causes the formation of a thin shell around the pattern. The thickness of the mold can be 10-20mm as compared to the heavy mold made for sand castings. When fully cured the skin is removed from the pattern, which is the shell mold.

For each ceramic shell molds there are two halves know as the cope and drag section. The two sections are joined by resin to form a complete shell mold. If an interior design is required, the cores are placed inside the mold before sealing the two parts.

For heavy castings, ceramic shell molds are held together by metals or other materials. Now, the molten metal is poured into the mold, and once it solidifies, the shell is broken to remove the casting. This process is highly useful for near net shape castings. Another advantage is that shell molding can be automated.

Automated Ceramic Shell Molding Machines and Robots: Shell molding machines like the cold shell molding machines helps in making castings with little molding material. In a cold shell molding machine the molds are made using cold binding materials. In it patterns made of wood, metal or plaster can be used. And the greatest benefit is that the mold can be kept horizontally or vertically.

Robotizing: Using robots for ceramic shell molding is a milestone for the old molding technology. Robots which are multi functional and re programmable are used in some foundries. Robots are used for a number of activities like robotic gate and sprue removal, robotic cutting of wedges for gate valves, robotic core setting, etc. The robots are reliable, consistent, more productive, provides better surface finish, and less machining etc.

Applications: A sizable amount of the casting in the steel industry are made by shell molding process, that ensures better profitability. Carbon steel, alloy steel, stainless steel, low alloys, aluminum alloys, copper, are all cast using shell molding process. Casting that require thin section and excellent dimensional accuracy are cast using this process. Body panes, truck hoods, small size boats, bath tubs, shells of drums, connecting rods, gear housings, lever arms, etc. are cast using croning process.

Advantages:

- Thin sections, complex parts and intricate designs can be cast.

- Excellent surface finish and good size tolerances.

- Less machining required for the castings.

- Near net shape castings, almost 'as cast' quality.

- Simplified process that can be handled by semi skilled operators.

- Full mechanized and automated casting process.

- Less foundry space required.

References

- Design-manual-1000000004-ceramics-and-ceramic-forming-techniques, resource: designophy.com, Retrieved 13 May, 2019

- Osborne, harold (ed), the oxford companion to the decorative arts, p. 746, 1975, oup, isbn 0198661134

- Deville, sylvain (april 2007). "ice-templated porous alumina structures". Acta materialia. 55 (6): 1965–1974. Arxiv:1710.04651. Doi:10.1016/j.actamat.2006.11.003

- Deville, sylvain (march 2008). "freeze-casting of porous ceramics: a review of current achievements and issues". Advanced engineering materials. 10 (3): 155–169. Arxiv:1710.04201. Doi:10.1002/adem.200700270

- Bigoni, d. Nonlinear solid mechanics: bifurcation theory and material instability. Cambridge university press, 2012 . Isbn 9781107025417

- Metal casting techniques - ceramic molding w.engineershandbook.com/mfgmethods/ceramic-molding.htm, archived from the original on 2007-10-25, retrieved 2010-12-15

- Degarmo, e. Paul; black, j t.; kohser, ronald a. (2003), materials and processes in manufacturing (9th ed.), wiley, pp. 315–316, isbn 0-471-65653-4

- Zoli, l.; sciti, d. (2017). "efficacy of a zrb 2 –sic matrix in protecting c fibres from oxidation in novel uhtcmc materials". Materials & design. 113: 207–213. Doi:10.1016/j.matdes.2016.09.104

- Casting-ceramic-shell-molds: industrialmetalcastings.com, Retrieved 6 January, 2019

4

Ceramography

The branch of science which is involved in creating objects from non-metallic, inorganic materials is known as ceramic engineering. It also deals with testing the fracture toughness, hardness and flexural strength of ceramics. The topics elaborated in this chapter will help in gaining a better perspective about these applications of ceramic engineering.

Ceramic engineering is the technology that involves the design and manufacture of ceramic products. Ceramics are inorganic, nonmetallic materials that have been hardened by baking at high temperatures. Highly regarded for being resistant to heat, these materials can be used for many demanding tasks that other materials, such as metals and polymers, cannot.

Until about the mid-twentieth century, the most important ceramics were the traditional clays, which were made into pottery, dinnerware, bricks, tiles, and decorative items. Since then, new materials called advanced ceramics have been prepared and are being used for a wide range of applications, including components used by the aerospace, automotive, defense, environmental, fiber-optic, and medical technologies. Ceramic parts are also used in cellular phones and personal computers. In addition, ceramic engineering is useful in such areas as petroleum refining, food processing, packaging, and mining.

Traditional ceramic raw materials include clay minerals such as kaolinite. Modern ceramic materials include silicon carbide and tungsten carbide, both of which are highly resistant to abrasion and are used in applications such as the wear plates of crushing equipment in mining operations. Each of NASA's Space Shuttles has a coating of ceramic tiles that protect it from the searing heat (up to 2,300 °F) produced during reentry into Earth's atmosphere. Thus, ceramic engineering is an important contributor to the modern technological revolution.

Properties of Ceramics

A ceramic material may be generally defined as any inorganic crystalline oxide material. It is solid and inert. Ceramic materials are brittle, hard, strong in compression, weak in shearing and tension. They withstand chemical erosion that occur in an acidic or caustic environment. In many cases withstanding erosion from the acid and bases applied to it. Ceramics generally can withstand very high temperatures such as temperatures that range from 1,000 °C to 1,600 °C (1,800 °F to 3,000 °F). Exceptions include inorganic materials that do not have oxygen such silicon carbide. Glass by definition is not a ceramic because it is an amorphous solid (non-crystalline). However, glass involves several steps of the ceramic process and its mechanical properties behave similarly to ceramic materials.

The Ceramic Process

The ceramic process generally follows this flow.

Milling → Batching → Mixing → Forming → Drying → Firing → Assembly

Milling is the process by which materials are reduced from a larger size to a smaller size. Milling may involve breaking up cemented material, thus the individual particle retain their shape or pulverization which involves grinding the particles themselves to a smaller size. Pulverization is actually fracturing the grains and breaking them down.

Generally, milling is done through mechanical means. The means include attrition which is particle to particle collision that results in agglomerate break up or particle shearing. Compression which is applying compressive forces that result in break up or fracturing. Another means is impact which involves a milling media—or the particles themselves—that cause break up or fracturing.

Examples of equipment that achieve attrition milling is a planetary mill or an wet attrition mill, also called wet scrubber. A wet scrubber is a machine that has paddles in water turning in opposite direction causing two vortexes turning into each other. The material in the vortex collide and break up.

Equipment that achieve compression milling include a jaw crusher, roller crusher, and cone crushers.

Finally, impact mills may include a ball mill with media that tumble and fracture material. Shaft impactors cause particle to particle attrition and compression which achieve size reduction.

Batching is the process of weighing the oxides according to recipes, and preparing them for mixing and drying.

Mixing occurs after batching and involve a variety of equipment such as dry mixing ribbon mixers (a type of cement mixer), Mueller mixers, and pug mills. Wet mixing generally involve the same equipment.

Forming is making the mixed material into shapes, ranging from toilet bowls to spark plug insulators. Forming can involve: Extrusion, such as extruding "slugs" to make bricks, Pressing to make shaped parts, or slip casting, as in making toilet bowls, wash basins and ornamentals like ceramic statues. Forming produces a "green" part, ready for drying. Green parts are soft, pliable, and over time will lose shape. Handling the green product product will change its shape. For example, a green brick can be "squeezed," and after squeezing it will stay that way.

Drying is removing the water or binder from the formed material. Spray drying is widely used to prepare powder for pressing operations. Other dryers are tunnel dryers and periodic dryers. Controlled heat is applied in this two-stage process. First, heat removes water. This step needs careful control, as rapid heating causes cracks and surface defects. The dried part is smaller than the green part, and is brittle, necessitating careful handling, since a small impact will cause crumbling and breaking.

Firing is where the dried parts pass through a controlled heating process, and the oxides are chemically changed to cause sintering and bonding. The fired part will be smaller than the dried part.

Assembly This process is for parts that require additional subassembly parts. In the case of a spark plug, the electrode is put into the insulator. This step does not apply to all ceramic products.

Applications

Ceramics can be used for many technological applications. One example is NASA's Space Shuttle, which uses ceramic tiles to protect it from the searing heat of reentry into Earth's atmosphere. Future supersonic space planes may likewise be fitted with such tiles. Ceramics are also used widely in electronics and optics. In addition to the applications listed here, ceramics are used as a coating in various engineering cases. For example, a ceramic bearing coating may be used over a titanium frame for an airplane. Recently, the field has come to include the studies of single crystals or glass fibers, in addition to traditional polycrystalline materials, and the applications of these have been overlapping and changing rapidly.

Aerospace

- Engines: Shielding a hot running airplane engine from damaging other components.

- Airframes: Used as a high-stress, high-temp and lightweight bearing and structural component.

- Missile nose-cones: Shielding the missile internals from heat.

- Space Shuttle tiles.

- Rocket Nozzles: Withstands and focuses the exhaust of the rocket booster.

Biomedical

- Artificial bone: Dentistry applications, teeth.

- Biodegradable splints: Reinforcing bones recovering from osteoporosis.

- Implant material.

Electronics and Electrical Industry

- Capacitors.

- Integrated Circuit packages.

- Transducers.

- Insulators.

Optical/Photonic

- Optical fibers; Glass fibers for super fast data transmission.

- Switches.

- Laser amplifiers.

- Lenses.

Modern-day Ceramic Engineering

Modern-day ceramic engineers may find themselves in a variety of industries, including mining and mineral processing, pharmaceuticals, foods, and chemical operations.

Now a multi-billion dollar a year industry, ceramics engineering and research has established itself as an important field of science. Applications continue to expand as researchers develop new kinds of ceramics to serve different purposes. An incredible number of ceramics engineering products have made their way into modern life.

CERAMOGRAPHY

Ceramography is the art and science of preparation, examination and evaluation of ceramic microstructures. Ceramography can be thought of as the metallography of ceramics. The microstructure is the structure level of approximately 0.1 to 100 μm, between the minimum wavelength of visible light and the resolution limit of the naked eye. The microstructure includes most grains, secondary phases, grain boundaries, pores, micro-cracks and hardness microindentions. Most bulk mechanical, optical, thermal, electrical and magnetic properties are significantly affected by the microstructure. The fabrication method and process conditions are generally indicated by the microstructure. The root cause of many ceramic failures is evident in the microstructure. Ceramography is part of the broader field of materialography, which includes all the microscopic techniques of material analysis, such as metallography, petrography and plastography. Ceramography is usually reserved for high-performance ceramics for industrial applications, such as 85–99.9% alumina (Al_2O_3) in figure, zirconia (ZrO_2), silicon carbide (SiC), silicon nitride (Si_3N_4), and ceramic-matrix composites. It is seldom used on whiteware ceramics such as sanitaryware, wall tiles and dishware.

Ceramographic Microstructures

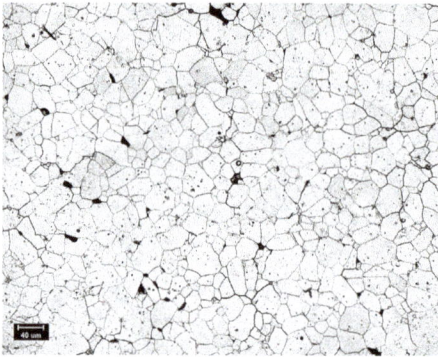

Thermally etched 99.9% alumina.

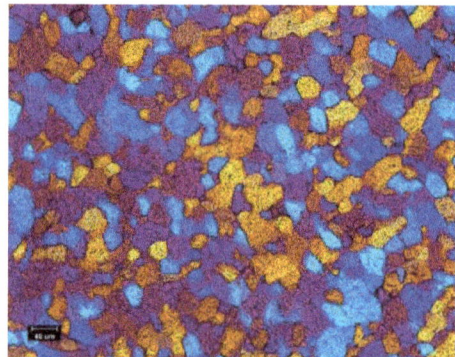

Thin section of 99.9% alumina.

Ceramography evolved along with other branches of materialography and ceramic engineering. Alois de Widmanstätten of Austria etched a meteorite in 1808 to reveal proeutectoid ferrite bands that grew on prior austenite grain boundaries. Geologist Henry Clifton Sorby, the "father of metallography," applied petrographic techniques to the steel industry in the 1860s in Sheffield, England. French geologist Auguste Michel-Lévy devised a chart that correlated the optical properties of minerals to their transmitted color and thickness in the 1880s. Swedish metallurgist J.A. Brinell invented the first quantitative hardness scale in 1900. Smith and Sandland developed the first microindention hardness test at Vickers Ltd. in London in 1922. Swiss-born microscopist A.I. Buehler started the first metallographic equipment manufacturer near Chicago in 1936. Frederick Knoop and colleagues at the National Bureau of Standards developed a less-penetrating (than Vickers) microindention test in 1939. Struers A/S of Copenhagen introduced the electrolytic polisher to metallography in 1943. George Kehl of Columbia University wrote a book that was considered the bible of materialography until the 1980s. Kehl co-founded a group within the Atomic Energy Commission that became the International Metallographic Society in 1967.

Preparation of Ceramographic Specimens

The preparation of ceramic specimens for microstructural analysis consists of five broad steps: Sawing, embedding, grinding, polishing and etching. The tools and consumables for ceramographic preparation are available worldwide from metallography equipment vendors and laboratory supply companies.

Sawing

Most ceramics are extremely hard and must be wet-sawed with a circular blade embedded with diamond particles. A metallography or lapidary saw equipped with a low-density diamond blade is usually suitable. The blade must be cooled by a continuous liquid spray.

Embedding

To facilitate further preparation, the sawed specimen is usually embedded (or mounted or encapsulated) in a plastic disc, 25, 30 or 35 mm in diameter. A thermosetting solid resin, activated by heat and compression, e.g. mineral-filled epoxy, is best for most applications. A castable (liquid) resin such as unfilled epoxy, acrylic or polyester may be used for porous refractory ceramics or microelectronic devices. The castable resins are also available with fluorescent dyes that aid in fluorescence microscopy. The left and right specimens in figure were embedded in mineral-filled epoxy. The center refractory in figure was embedded in castable, transparent acrylic.

Grinding

Grinding is abrasion of the surface of interest by abrasive particles, usually diamond, that are bonded to paper or a metal disc. Grinding erases saw marks, coarsely smooths the surface, and removes stock to a desired depth. A typical grinding sequence for ceramics is one minute on a 240-grit metal-bonded diamond wheel rotating at 240 rpm and lubricated by flowing water, followed by a similar treatment on a 400-grit wheel. The specimen is washed in an ultrasonic bath after each step.

Polishing

Polishing is abrasion by free abrasives that are suspended in a lubricant and can roll or slide between the specimen and paper. Polishing erases grinding marks and smooths the specimen to a mirror-like finish. Polishing on a bare metallic platen is called lapping. A typical polishing sequence for ceramics is 5–10 minutes each on 15-, 6- and 1-μm diamond paste or slurry on napless paper rotating at 240 rpm. The specimen is again washed in an ultrasonic bath after each step. The three sets of specimens in Fig. have been sawed, embedded, ground and polished.

Etching

Etching reveals and delineates grain boundaries and other microstructural features that are not apparent on the as-polished surface. The two most common types of etching in ceramography are selective chemical corrosion, and a thermal treatment that causes relief. As an example, alumina can be chemically etched by immersion in boiling concentrated phosphoric acid for 30–60 s, or thermally etched in a furnace for 20–40 min at 1,500 °C (2,730 °F) in air. The plastic encapsulation must be removed before thermal etching. The alumina in figure was thermally etched.

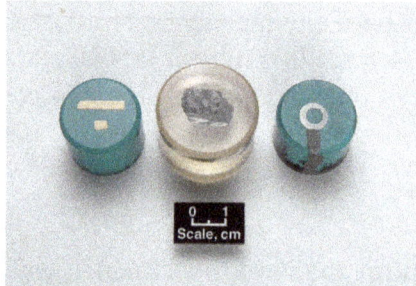

Embedded, polished ceramographic sections.

Alternatively, non-cubic ceramics can be prepared as thin sections, also known as petrography, for examination by polarized transmitted light microscopy. In this technique, the specimen is sawed to ~1 mm thick, glued to a microscope slide, and ground or sawed (e.g., by microtome) to a thickness (x) approaching 30 μm. A cover slip is glued onto the exposed surface. The adhesives, such as epoxy or Canada balsam resin, must have approximately the same refractive index ($\eta \approx 1.54$) as glass. Most ceramics have a very small absorption coefficient in the Beer-Lambert law below, and can be viewed in transmitted light. Cubic ceramics, e.g. yttria-stabilized zirconia and spinel, have the same refractive index in all crystallographic directions and appear, therefore, black when the microscope's polarizer is 90° out of phase with its analyzer.

$$I_t = I_0 e^{-\alpha x} \text{ (Beer–Lambert eqn)}$$

Ceramographic specimens are electrical insulators in most cases, and must be coated with a conductive ~10-nm layer of metal or carbon for electron microscopy, after polishing and etching. Gold or Au-Pd alloy from a sputter coater or evaporative coater also improves the reflection of visible light from the polished surface under a microscope, by the Fresnel formula below. Bare alumina ($\eta \approx 1.77$, $k \approx 10^{-6}$) has a negligible extinction coefficient and reflects only 8% of the incident light from the microscope, as in figure. Gold-coated ($\eta \approx 0.82$, $k \approx 1.59$ @ $\lambda = 500$ nm) alumina reflects 44% in air, 39% in immersion oil.

$$R = \frac{I_r}{I_i} = \frac{(\eta_1 - \eta_2)^2 + k^2}{(\eta_1 + \eta_2)^2 + k^2} \text{ (Fresnel eqn)}$$

Ceramographic Analysis

Ceramic microstructures are most often analyzed by reflected visible-light microscopy in brightfield. Darkfield is used in limited circumstances, e.g., to reveal cracks. Polarized transmitted light is used with thin sections, where the contrast between grains comes from birefringence. Very fine microstructures may require the higher magnification and resolution of a scanning electron microscope (SEM) or confocal laser scanning microscope (CLSM). The cathodoluminescence microscope (CLM) is useful for distinguishing phases of refractories. The transmission electron microscope (TEM) and scanning acoustic microscope (SAM) have specialty applications in ceramography.

Ceramography is often done qualitatively, for comparison of the microstructure of a component to a standard for quality control or failure analysis purposes. Three common quantitative analyses of microstructures are grain size, second-phase content and porosity. Microstructures are measured by the principles of stereology, in which three-dimensional objects are evaluated in 2-D by projections or cross-sections.

Grain size can be measured by the line-fraction or area-fraction methods of ASTM E112. In the line-fraction methods, a statistical grain size is calculated from the number of grains or grain boundaries intersecting a line of known length or circle of known circumference. In the area-fraction method, the grain size is calculated from the number of grains inside a known area. In each case, the measurement is affected by secondary phases, porosity, preferred orientation, exponential distribution of sizes, and non-equiaxed grains. Image analysis can measure the shape factors of individual grains by ASTM E1382.

Second-phase content and porosity are measured the same way in a microstructure, such as ASTM E562. Procedure E562 is a point-fraction method based on the stereological principle of point fraction = volume fraction, i.e., $P_p = V_v$. Second-phase content in ceramics, such as carbide whiskers in an oxide matrix, is usually expressed as a mass fraction. Volume fractions can be converted to mass fractions if the density of each phase is known. Image analysis can measure porosity, pore-size distribution and volume fractions of secondary phases by ASTM E1245. Porosity measurements do not require etching. Multi-phase microstructures do not require etching if the contrast between phases is adequate, as is usually the case.

Grain size, porosity and second-phase content have all been correlated with ceramic properties such as mechanical strength σ by the Hall–Petch equation. Hardness, toughness, dielectric constant and many other properties are microstructure-dependent.

Microindention Hardness and Toughness

The hardness of a material can be measured in many ways. The Knoop hardness test, a method of microindention hardness, is the most reproducible for dense ceramics. The Vickers hardness test and superficial Rockwell scales (e.g., 45N) can also be used, but tend to cause more surface damage than Knoop. The Brinell test is suitable for ductile metals, but not ceramics. In the Knoop test, a diamond indenter in the shape of an elongated pyramid is forced into a polished (but not etched) surface under a predetermined load, typically 500 or 1000 g. The load is held for some amount of time, say 10 s, and the indenter is retracted. The indention long diagonal is measured under a microscope, and the Knoop hardness (HK) is calculated from the load (P, g) and the square of the diagonal length in the equations below. The constants account for the projected area of the indenter and unit conversion factors. Most oxide ceramics have a Knoop hardness in the range of 1000–1500 kg$_f$/mm^2 (10 – 15 GPa), and many carbides are over 2000 (20 GPa). The method is specified in ASTM C849, C1326 & E384. Microindention hardness is also called microindentation hardness or simply microhardness. The hardness of very small particles and thin films of ceramics, on the order of 100 nm, can be measured by nanoindentation methods that use a Berkovich indenter.

$$HK = 14229\frac{P}{d^2}\ (\text{kg}_f/\text{mm}^2)\ \text{and}\ HK = 139.54\frac{P}{d^2}\ (\text{GPa})$$

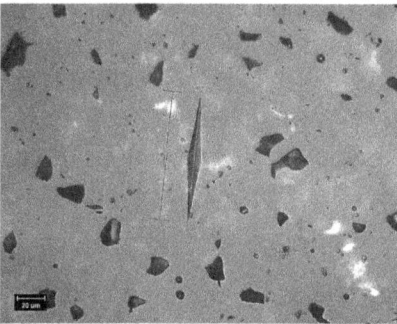
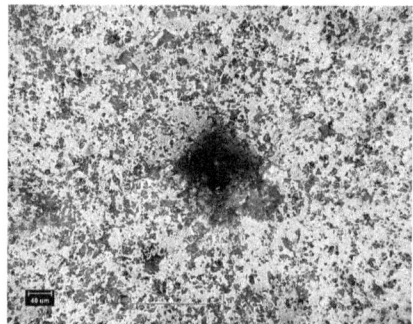

Knoop indention (P=1kg) in 99.5% alumina.   Toughness indention (P=10kg) in 96% alumina.

The toughness of ceramics can be determined from a Vickers test under a load of 10 – 20 kg. Toughness is the ability of a material to resist crack propagation. Several calculations have been formulated from the load (P), elastic modulus (E), microindention hardness (H), crack length and flexural strength (σ). Modulus of rupture (MOR) bars with a rectangular cross-section are indented in three places on a polished surface. The bars are loaded in 4-point bending with the polished, indented surface in tension, until fracture. The fracture normally originates at one of the indentions. The crack lengths are measured under a microscope. The toughness of most ceramics is 2–4 MPa√m, but toughened zirconia is as much as 13, and cemented carbides are often over 20. The toughness-by-indention methods have been discredited recently and are being replaced by more rigorous methods that measure crack growth in a notched beam in flexure.

$$K_{icl} = 0.016\sqrt{\frac{E}{H}}\frac{P}{(c_0)^{1.5}}\ \text{initial crack length,}$$

$$K_{isb} = 0.59\left(\frac{E}{H}\right)^{1/8}[\sigma(P^{1/3})]^{3/4}\ \text{indention strength in bending.}$$

NANOPHASE CERAMICS

Nanophase ceramics are ceramics that are nanophase materials (that is, materials that have grain sizes under 100 nanometers). They have the potential for superplastic deformation. Because of the small grain size and added grain boundaries properties such as ductility, hardness, and reactivity see drastic changes from ceramics with larger grains.

Structure of Nanophase Ceramics

The structure of nanophase ceramics is not too different than that of ceramics. The main difference is the amount of surface area per mass. Particles of ceramics have small surface areas, but when those particles are shrunk to within a few nanometers, the surface area of the same amount of a mass of a ceramic greatly increases. So in general, nanophase materials have greater surface areas than that of a similar mass material at a larger scale. This is important because if the surface area is very large the particles can be in contact with more of their surroundings, which in turn increases the reactivity of the material. The reactivity of a material changes the material's mechanical properties and chemical properties, among many other things. This is especially true in nanophase ceramics.

Properties of Nanophase Ceramics

Nanophase ceramics have unique properties than regular ceramics due to their improved reactivity. Nanophase ceramics exhibit different mechanical properties than their counterpart such as higher hardness, higher fracture toughness, and high ductility. These properties are far from ceramics which behave as brittle, low ductile materials.

Titanium Dioxide

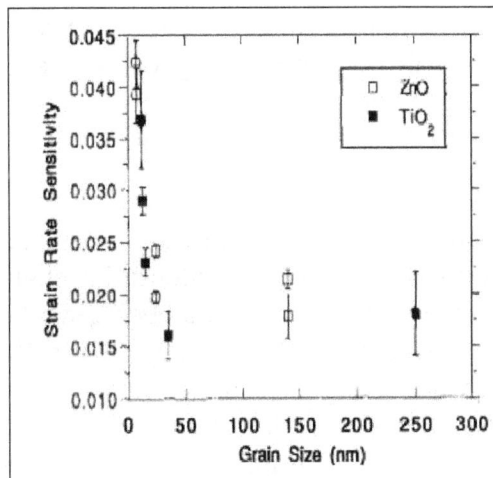

Strain rate sensitivity for TiO_2.

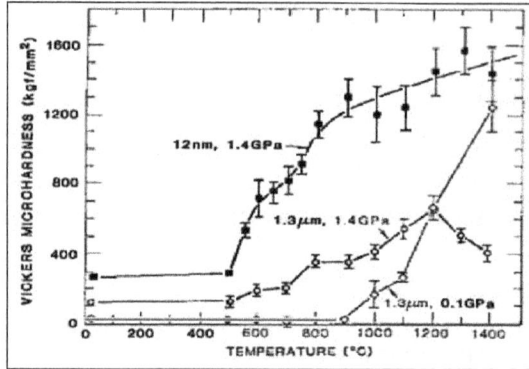

Microhardness of TiO_2.

Titanium dioxide (TiO_2), has been shown to have increased hardness and ductility at the nanoscale. In an experiment, grains of titanium dioxide that had an average size of 12 nanometers were compressed at 1.4 GPa and sintered at 200 °C. The result was a grain hardness of about 2.2 times greater than that of grains of titanium dioxide with an average size of 1.3 micrometers at the same temperature and pressure. In the same experiment, the ductility of titanium dioxide was measured. The strain rate sensitivity of a 250 nanometer grain of titanium dioxide was about 0.0175, while a grain with size of about 20 nanometers had a strain rate sensitivity of approximately .037; a significant increase.

Processing of Nanophase Ceramics

Nanophase ceramics can be processed from atomic, molecular, or bulk precursors. Gas condensation, chemical precipitation, aerosol reactions, biological templating, chemical vapor deposition, and physical vapor deposition are techniques used to synthesis nanophase ceramics from molecular or atomic precursors. To process nanophase ceramics from bulk precursors, mechanical attrition, crystallization from the amorphous state, and phase separation are used to create nanophase ceramics. Synthesizing nanophase ceramics from atomic or molecular precursors are desired more because a greater control over microscopic aspects of the nanophase ceramic can occur.

Gas Condensation

Gas condensation is one way nanophase ceramics are produced. First, precursor ceramics are evaporated from sources within a gas-condensation chamber. Then the ceramics are condensed in a gas (dependent on the material being synthesized) and transported via convection to a liquid-nitrogen filled cold finger. Next, the ceramic powders are scraped off the cold finger and collect in a funnel below the cold finger. The ceramic powders then become consolidated in a low-pressure compaction device and then in a high-pressure compaction device. This all occurs in a vacuum, so no impurities can enter the chamber and affect the results of the nanophase ceramics.

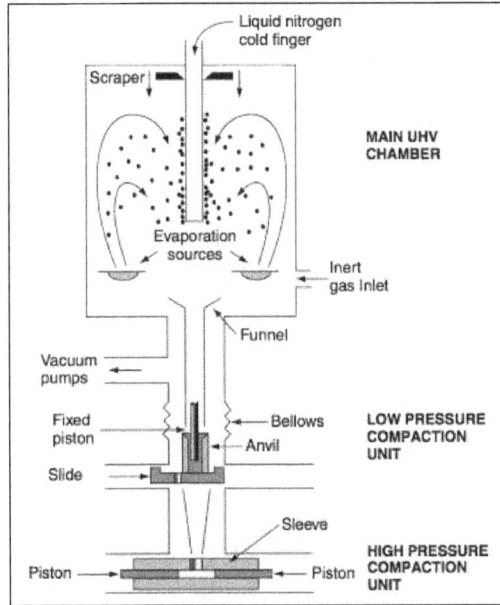

Synthesis of nanophase ceramics using gas condensation.

Applications

Nanophase ceramics have unique properties that make them optimal for a variety of applications.

Drug Delivery

Materials used in drug delivery in the past ten years have primarily been polymers. However, nanotechnology has opened the door for the use of ceramics with benefits not previously seen in polymers. The large surface area to volume ratio of nanophase materials makes it possible for large amounts of drugs to be released over long periods of time. Nanoparticles to be filled with drugs can be easily manipulated in size and composition to allow for increased endocytosis of drugs into targeted cells and increased dispersion through fenestrations in capillaries. While these benefits all relate to nanoparticles in general (including polymers), ceramics have other, unique abilities. Unlike polymers, slow degradation of ceramics allows for longer release of the drug. Polymers also tend to swell in liquid which can cause an unwanted burst of drugs. The lack of swelling shown by most ceramics allows for increased control. Ceramics can also be created to match the chemistry of biological cells in the body increasing bioactivity and biocompatibility. Nanophase ceramic drug carriers are also able to target specific cells. This can be done by manufacturing a material to bond to the specific cell or by applying an external magnetic field, attracting the carrier to a specific location.

Bone Substitution

Nanophase ceramics have great potential for use in orthopedic medicine. Bone and collagen have structures on the nanoscale. Nanomaterials can be manufactured to simulate these structures which is necessary for grafts and implants to successfully adapt to and handle varying stresses. The surface properties of nanophase ceramics is also very important for bone substitution and regeneration. Nanophase ceramics have much rougher surfaces than larger materials and also have increased surface area. This promotes reactivity and absorption of proteins that assist tissue development. Nano-hydroxyapatite is one nanophase ceramic that is used as a bone substitute. Nano grain size increases the bonding, growth, and differentiation of osteoblasts onto the ceramic. The surfaces of nanophase ceramics can also be modified to be porous allowing osteoblasts to create bone within the structure. The degradation of the ceramic is also important because the rate can be changed by changing the crystallinity. This way as bone grows the substitute can diminish at a similar rate.

FRACTURE TOUGHNESS TESTS OF CERAMICS

Fracture is a process of breaking a solid into pieces as a result of stress.

There are two principal stages of the fracture process:

- Crack formation,
- Crack propagation.

There are two fracture mechanisms: Ductile fracture and brittle fracture.

Ceramic materials have extremely low ductility, therefore they failure by brittle mechanism.

- Brittle fracture,
- Fracture Toughness,
- Flexure Test,
- Indentation Fracture Test.

Brittle Fracture

Brittle fracture is characterized by very low Plastic deformation and low energy absorption prior to breaking. A crack, formed as a result of the brittle fracture, propagates fast and without increase of the stress applied to the material. The brittle crack is perpendicular to the stress direction.

There are two possible mechanisms of the brittle fracture: transcrystalline (transgranular, cleavage) or intercrystalline (intergranular). Cleavage cracks pass along crystallographic planes through the grains.

Intercrystalline fracture occurs through the grain boundaries, embrittled by segregated impurities, second phase inclusions and other defects.

The brittle fractures usually possess bright granular appearance.

Fracture Toughness

Fracture Toughness is ability of material to resist fracture when a crack is present.

The general factors, affecting the fracture toughness of a material are: Temperature, strain rate, presence of structure defects and presence of stress concentration (notch) on the specimen surface.

Stress-intensity Factor (K) is a quantitative parameter of fracture toughness determining a maximum value of stress which may be applied to a specimen containing a crack (notch) of a certain length.

Depending on the direction of the specimen loading and the specimen thickness, four types of stress-intensity factors are used: K_C, K_{IC}, K_{IIC}, K_{IIIC}.

- K_C– stress-intensity factor of a specimen, thickness of which is less than a critical value.

  KC depends on the specimen thickness. This condition is called plane stress.

- K_{IC}, K_{IIC}, K_{IIIC}– stress-intensity factors, relating to the specimens, thickness of which is above the critical value therefore the values of K_{IC}, K_{IIC}, K_{IIIC} do not depend on the specimen thickness. This condition is called plane strain.

- K_{IIC} and K_{IIIC} – stress-intensity factors relating to the fracture modes in which the loading direction is parallel to the crack plane. These factors are rarely used for metallic materials and are not used for ceramics.

- K_{IC} – plane strain stress-intensity factor relating to the fracture modes in which the loading direction is normal to the crack plane. This factor is widely used for both metallic and ceramic materials.

K_{IC} is used for estimation critical stress applied to a specimen with a given crack length:

$$σC ≤ K_{IC} / (Y(π\,a)^{1/2})$$

Where:

K_{IC} – stress-intensity factor, measured in MPa*m$^{1/2}$.

σC– the critical stress applied to the specimen.

a – the crack length for edge crack or half crack length for internal crack.

Y – geometry factor.

Two test methods are used for measuring fracture toughness parameter (stress-intensity factor) of ceramic materials: Flexure Test and Indentation Fracture Test.

Flexure Test

The test method is similar to that which is used for measuring Flexural Strength, however notched specimens are used.

Indentation Fracture Test

Vickers Hardness Method is used for this test.

Polished surface of a ceramic sample is indented by Vickers Indenter, resulting in formation of four cracks emanating from the indent corners.

The cracks length is inversely proportion to the material toughness; therefore KIC may be estimated by measuring the cracks length.

CERAMIC COATINGS FOR METAL PARTS

There are a wide range of ceramic coating materials that can be applied to metal components in order to enhance their functional properties. Most ceramic coatings are electrically nonconductive (making them excellent insulators), have a significantly higher level of abrasion resistance than most metals, and are capable of maintaining their integrity under severely elevated temperatures, sometimes up to 4,500 degrees Fahrenheit. Wear-resistant ceramics, such as titanium nitride and chromium carbide, can be applied to work steels and air-hardening tool steels via chemical vapor deposition (CVD), which is one of the more common application methods currently in use.

Before applying a coating, it is important to ensure compatibility between the ceramic material and the metal surface. Much of this compatibility depends on thermal expansion properties, as having expansion rates that differ too greatly can cause a coating to crack when it is being cooled after application. In addition, a diffusion layer typically forms on the metal surface, and this can lead to a coating that is too soft or too brittle for the design specifications. Complementary thermal properties will help to prevent heat checking and improve resistance to wear and fracture.

Ceramic Coating Manufacturing Applications

Ceramic coatings are often used as barrier materials to enhance the interaction between moving metal parts, such as in the automotive industry. However, they are also increasingly being employed to augment certain manufacturing processes, and exhibit potential for improving the efficiency of some fabricating methods. Ceramic coatings are sturdy and have a high level of lubricity, but due to oxidation concerns, they are typically used in temperatures under 1,200 degrees (F). However, this allows them to be applied to hot forging dies, which operate at lower temperatures. Ceramic coating increases the operational lifespan for these dies, allowing them to produce a greater number of parts before wearing down. Ceramic materials, such as magnesium zirconate and zirconia, exhibiting a high level of hardness, thermal resistance, and elevated melting points are being used as heat barrier coatings for industrial parts.

Ceramic Coating Processes

Applying a ceramic coating to a substrate is multi-stage process. The preparatory phases of cleaning, roughening, and undercoating (or priming) greatly influence the success of the project. The actual coating effectiveness depends largely on the mechanical, chemical, and physical bonds that determine the coating adherence and ultimate strength of the ceramic layer. Aside from chemical vapor deposition, the most common ceramic coating methods include:

- Plasma Spraying: In plasma spraying, ceramic powder is passed through an ionized gas at extremely high temperatures, sometimes approaching 30,000 degrees (F). The pressurized gas speeds molten ceramic particles toward the substrate where they bond onto its surface. The result is a strongly-adhering and high-density coating, but the process can be very expensive.

- Detonation Gun: The detonation gun process is most effective for particular ceramic materials, such a tungsten carbide, that are required for producing highly dense coatings on a metal surface. It creates an explosion of oxygen and acetylene gas at around 6,000 degrees (F), melting the ceramic and firing it at high speed toward the target substrate.

- Oxygen Acetylene Powder: This method involves heating ceramic powder under a 5000 degree (F) flame, and using compressed gas to spray the coating onto the substrate. It creates porous coating layers with relatively low adhesion strength.

- Oxygen Acetylene Rod: In this method, a fused ceramic rod is passed under an oxyacetylene torch burning at 500 degrees (F). Pressurized gas is then used to spray molten ceramic material onto a surface, producing a coating with a high level of cohesive bonding.

In addition to these standard processing methods, continuing research in ceramic coating technology has introduced newer techniques that may have a major influence on future ceramics work. For example, a procedure for coating metalworking dies with refractory materials, such as molybdenum and tungsten, employs a plasma spray gun and low-shearing compaction to achieve a highly effective and wear-resistant coating.

HARDNESS TESTING OF CERAMIC COATINGS

Ceramic coatings are used in a wide variety of industrial and commercial manufacturing processes, including within the automotive industry, in manufacturing processes, and in the design of industrial parts. These coatings are often used to make mechanical components harder, more resistant to corrosion (i.e. rust), impervious to liquids and gases, better insulated (both electrically and thermally), and overall more resistant to wear.

Selecting a Ceramic Coating

The type of ceramic coating and appropriate method of deposition depends on various factors:

- The desired thickness of the coating.

- The desired function of the ceramic coating.

- The expectations of the component's operation.

- Various economic considerations.

- The substrate material.

- The size and shape of the area to be coated.

Thin coatings, such as those used on cutting tools, are most often applied through a vapor deposition process, either chemical (CVD) or physical (PVD). Thick coatings, such as those used in decorative enamels, are commonly applied by a thermal spraying and enameling process.

Types of ceramic coatings include titanium nitride and chromium carbide. These coatings can be applied to steel and other metal parts to increase resistance to abrasion, and to increase a material's ability to withstand extremely high temperatures.

The Purpose of Hardness Testing for Ceramic Coatings

The mechanical property of a material's hardness is defined as a specific and calculable measure of how resistant a material is to compressive force. Hardness is a representation of a material's resistance to localised plastic deformation.

Strong intermolecular bonds generally produce a high level of hardness in a material. The measurement is dependent on the measured elasticity, ductility, stiffness, plasticity, strain, strength, viscosity, and toughness of the material being evaluated for hardness.

Applications of hardness testing of ceramics include defining the hardness gradient of a material, the surface hardness, the case depth, coating hardness, phase hardness, grain hardness, hardness at grain boundaries, and hardness of powders.

The hardness of a ceramic is defined by its chemical composition, including porosity, grain size, and grain-boundary phases. Ceramic hardness is usually tested using either the Vickers or Knoop method, most often using diamond indenters.

The Methods and Process of Hardness Testing

There are multiple measurements of hardness, including scratch hardness, indentation hardness, and rebound hardness. Each type of measurement is based on an individual measurement scale, however, conversion between scales is possible for practical purposes.

Indentation tests are a common method of testing the hardness of a ceramic material. Indentation is a straight forward test of penetrating a given material with an indenter under a pre-defined indentation load, then measuring the resulting indentation.

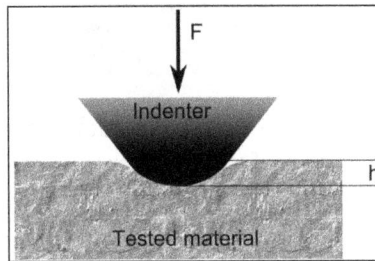

Indenters come in a variety of different shapes and sizes, and the load can be set for nano, micro, or macro indentation ranges, so as to specify the range of mechanical properties that will be tested.

While direct comparison across the various methods of hardness testing is not always possible, the general concept behind the measurement is very similar: the harder the testing material is, the smaller the indentation will be.

Dense ceramics are often measured for hardness by the Knoop hardness test, a method of microindentation optimized for brittle materials or thin coatings such as ceramic. The Knoop hardness test is most practical for the purpose of ceramic coating tests, as only a small indentation is required to evaluate and measure a material's hardness. Other methods, such as the Vickers hardness test and Rockwell scales, can also be used to determine hardness, but are known to cause more damage to the testing material than the Knoop method.

In the Knoop test, a diamond indenter shaped like a long pyramid is used on a polished surface under a predetermined load for a predetermined length of time (an example of both is a 500 g load held for 10 seconds). The indention is then measured under a microscope and the Knoop hardness (HK) is calculated. The formula used to determine the Knoop hardness of a material is as follows:

$$HK = \text{load (kgf)/impression area (mm}^2)$$

$$= P/C_p L^2$$

Where P equals the load, C_p equals the correction factor related to the shape of the indenter (generally 0.070279), and L equals the length of the indentation along its long axis. The majority of oxide ceramics tested have a Knoop hardness of 1000 to 1500 kg_f/mm^2.

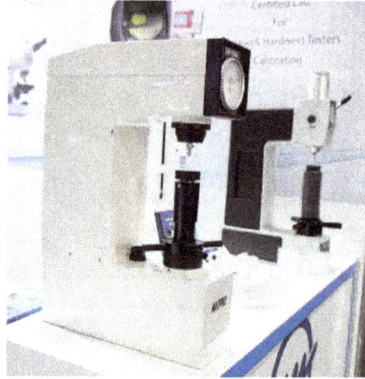

The Knoop method has a few disadvantages; particularly worth noting is the need to use a microscope to measure the size of the indentation created by the test. The amount of time required to apply the indenter may also be considered a drawback of this test.

The Vickers hardness test method uses a square-based diamond pyramid indenter to penetrate the testing material.

The load, generally weighted between 1 and 120 kgf, is applied for 30 seconds and calculated using the following formula:

$$HV = 1.854 * F/D^2$$

Where F is the measurement of the applied load in kg and D is the length of the impression diagonal in mm.

As in the Knoop test, the impression is measured on the diagonal with the assistance of a microscope. The Vickers four-sided indenter is known to crack brittle materials, in which case the Knoop method is likely preferred. It is also worth noting that a direct comparison between Vickers and Knoop hardness numbers is not possible due to the differences in indentation method.

Thinner ceramic coatings are often measured by nanoindentation methods using a Berkovich indenter. The Berkovich tip is a nearly flat, three-sided pyramid with a sharp point used to make indentations to test the hardness of materials greater than 100 nanometers thick.

Similar to the Knoop method, nanoindentation requires the placement of an indenter tip, such as the Berkovich indenter in this case, resulting in the measurement of the indentation created by the added pressure of a defined load. In this scenario, hardness (or H) is equal to the max load (or P_{max}) over A_r (or the residual indentation area).

FLEXURAL STRENGTH TESTS OF CERAMICS

Extremely low ductility of ceramic materials does not allow measuring their mechanical properties by conventional tensile test, which is widely used for metals.

Brittle Materials, including ceramics, are tested by Flexure Test (Transverse Beam Test, Bending Test).

There are two standard Flexure Test methods:

- 3-point Flexure Test.

- 4-point Flexure Test.

- Flexural strength calculation.

3 Point Flexure Test

In this test a specimen with round, rectangular or flat cross-section is placed on two parallel supporting pins. The loading force is applied in the middle by means loading pin.

The supporting and loading pins are mounted in a way, allowing their free rotation about:

- Axis parallel to the pin axis.

- Axis parallel to the specimen axis.

This configuration provides uniform loading of the specimen and prevents friction between the specimen and the supporting pins.

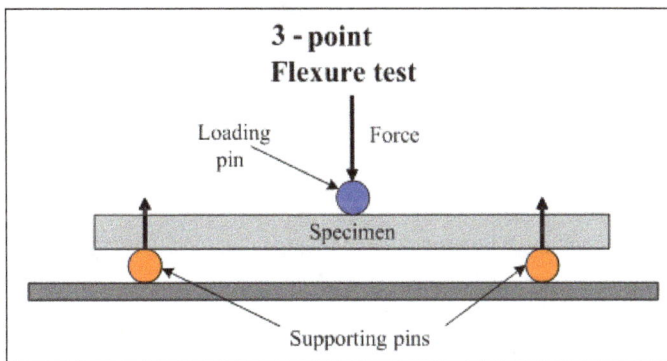

4 Point Flexure Test

In this test the loading force is applied by means of two loading pins with a distance between them equal to a half of the distance between the supporting pins.

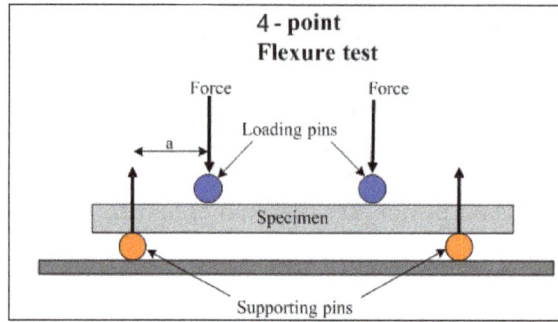

Flexural Strength Calculation

As a result of the loading, the specimen bends, causing formation of in its convex side and compression stress in the concave side.

The cross head speed in flexural test normally varies within the range 0.004-0.4 inch/min (0.1-10 mm/min). Speeds 1 mm/min and 0.1 "/min (2.54 mm/min) are mostly used in the tests.

The maximum stress and corresponding maximum strain are calculated for every load value.

The results are then plotted in the stress-strain diagram.

Modulus of Rupture (Flexural Strength) is the stress of the extreme fiber of a specimen at its failure in the Flexure Test.

Flexural Strength is calculated by the formula:

$\sigma = 3LF/(2bd^2)$ in 3-point test of rectangular specimen.

$\sigma = 3Fa/(bd^2)$ in 4-point test of rectangular specimen.

$\sigma = 16Fa/(\pi D^3) = 2Fa/(\pi r^3)$ in 4-point test of round specimen.

Where:

L – Specimen length.

F – Total force applied to the specimen by two loading pins.

b – Specimen width.

d – Specimen thickness.

r – Specimen section radius.

a - Distance between the supporting and loading pins.

D – Section diameter of round specimen.

BIOCERAMICS

A porous bioceramic granule of an orthobiologic
calcium composition.

Bioceramics and bioglasses are ceramic materials that are biocompatible. Bioceramics are an important subset of biomaterials. Bioceramics range in biocompatibility from the ceramic oxides, which are inert in the body, to the other extreme of resorbable materials, which are eventually replaced by the body after they have assisted repair. Bioceramics are used in many types of medical procedures. Bioceramics are typically used as rigid materials in surgical implants, though some bioceramics are flexible. The ceramic materials used are not the same as porcelain type ceramic materials. Rather, bioceramics are closely related to either the body's own materials or are extremely durable metal oxides.

Applications

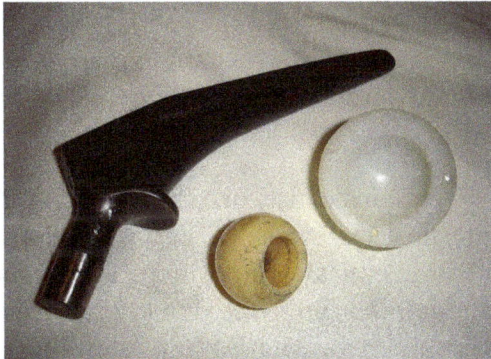

A titanium hip prosthesis, with a ceramic
head and polyethylene acetabular cup.

Ceramics are now commonly used in the medical fields as dental and bone implants. Surgical cermets are used regularly. Joint replacements are commonly coated with bioceramic materials to reduce wear and inflammatory response. Other examples of medical uses for bioceramics are in pacemakers, kidney dialysis machines, and respirators. The global demand on medical ceramics and ceramic components was about U.S. $9.8 billion in 2010. It was forecast to have an annual growth of 6 to 7 percent in the following years, with world market value predicted to increase to U.S. $15.3 billion by 2015 and reach U.S. $18.5 billion by 2018.

Mechanical Properties and Composition

Bioceramics are meant to be used in extracorporeal circulation systems (dialysis for example) or engineered bioreactors; however, they're most common as implants. Ceramics show numerous applications as biomaterials due to their physico-chemical properties. They have the advantage of being inert in the human body, and their hardness and resistance to abrasion makes them useful for bones and teeth replacement. Some ceramics also have excellent resistance to friction, making them useful as replacement materials for malfunctioning joints. Properties such as appearance and electrical insulation are also a concern for specific biomedical applications.

Some bioceramics incorporate alumina (Al_2O_3) as their lifespan is longer than that of the patient's. The material can be used in inner ear ossicles, ocular prostheses, electrical insulation for pacemakers, catheter orifices and in numerous prototypes of implantable systems such as cardiac pumps.

Aluminosilicates are commonly used in dental prostheses, pure or in ceramic-polymer composites. The ceramic-polymer composites are a potential way to filling of cavities replacing amalgams suspected to have toxic effects. The aluminosilicates also have a glassy structure. Contrary to artificial teeth in resin, the colour of tooth ceramic remains stable Zirconia doped with yttrium oxide has been proposed as a substitute for alumina for osteoarticular prostheses. The main advantages are a greater failure strength, and a good resistance to fatigue.

Vitreous carbon is also used as it is light, resistant to wear, and compatible with blood. It is mostly used in cardiac valve replacement. Diamond can be used for the same application, but in coating form.

Calcium phosphate-based ceramics constitute, at present, the preferred bone substitute in orthopaedic and maxillofacial surgery. They are similar to the mineral phase of the bone in structure and/or chemical composition. The material is typically porous, which provide a good bone-implant interface due to the increase of surface area that encourages cell colonisation and revascularisation. Additionally, it has lower mechanical strength compared to bone, making highly porous implants very delicate. Since Young's modulus of ceramics is generally much higher than that of the bone tissue, the implant can cause mechanical stresses at the bone interface. Calcium phosphates usually found in bioceramics include hydroxyapatite (HAP) $Ca_{10}(PO_4)_6(OH)_2$; tricalcium phosphate β (β TCP): $Ca_3(PO_4)_2$; and mixtures of HAP and β TCP.

Table: Bioceramics Applications

Devices	Function	Biomaterial
Artificial total hip, knee, shoulder, elbow, wrist	Reconstruct arthritic or fractured joints	High-density alumina, metal bioglass coatings
Bone plates, screws, wires	Repair fractures	Bioglass-metal fibre composite, Polysulphone-carbon fibre composite
Intramedullary nails	Align fractures	Bioglass-metal fibre composite, Polysulphone-carbon fibre composite
Harrington rods	Correct chronic spinal curvature	Bioglass-metal fibre composite, Polysulphone-carbon fibre composite
Permanently implanted artificial limbs	Replace missing extremities	Bioglass-metal fibre composite, Polysulphone-carbon fibre composite
Vertebrae Spacers and extensors	Correct congenital deformity	Al_2O_3
Spinal fusion	Immobilise vertebrae to protect spinal cord	Bioglass
Alveolar bone replacements, mandibular reconstruction	Restore the alveolar ridge to improve denture fit	Polytetra fluro ethylene (PTFE) - carbon composite, Porous Al_2O_3, Bioglass, dense-apatite
End osseous tooth replacement implants	Replace diseased, damaged or loosened teeth	Al_2O_3, Bioglass, dense hydroxyapatite, vitreous carbon
Orthodontic anchors	Provide posts for stress application required to change deformities	Bioglass-coated Al_2O_3, Bioglass coated vitallium

Table: Mechanical Properties of Ceramic Biomaterials.

Material	Young's Modulus (GPa)	Compressive Strength (MPa)	Bond strength (GPa)	Hardness	Density (g/cm³)
Inert Al_2O_3	380	4000	300-400	2000-3000(HV)	>3.9
ZrO_2 (PS)	150-200	2000	200-500	1000-3000(HV)	≈6.0
Graphite	20-25	138	NA	NA	1.5-1.9
(LTI)Pyrolitic Carbon	17-28	900	270-500	NA	1.7-2.2
Vitreous Carbon	24-31	172	70-207	150-200(DPH)	1.4-1.6
Bioactive HAP	73-117	600	120	350	3.1
Bioglass	≈75	1000	50	NA	2.5
AW Glass Ceramic	118	1080	215	680	2.8
Bone	3-30	130-180	60-160	NA	NA

Multipurpose

A number of implanted ceramics have not actually been designed for specific biomedical applications. However, they manage to find their way into different implantable systems because of their properties and their good biocompatibility. Among these ceramics, we can cite silicon carbide, titanium nitrides and carbides, and boron nitride. TiN has been suggested as the friction surface in hip prostheses. While cell culture tests show a good biocompatibility, the analysis of implants shows significant wear, related to a delaminating of the TiN layer. Silicon carbide is another modern-day ceramic which seems to provide good biocompatibility and can be used in bone implants.

Specific use

In addition to being used for their traditional properties, bioactive ceramics have seen specific use for due to their biological activity. Calcium phosphates, oxides, and hydroxides are common examples. Other natural materials — generally of animal origin — such as bioglass and other composites feature a combination of mineral-organic composite materials such as HAP, alumina, or titanium dioxide with the biocompatible polymers (polymethylmethacrylate): PMMA, poly(L-lactic) acid: PLLA, poly(ethylene). Composites can be differentiated as bioresorbable or non-bioresorbable, with the latter being the result of the combination of a non-bioresorbable calcium phosphate (HAP) with a non-bioresorbable polymer (PMMA, PE). These materials may become more widespread in the future, on account of the many combination possibilities and their aptitude at combining a biological activity with mechanical properties similar to those of the bone.

Biocompatibility

Bioceramics' properties of being anticorrosive, biocompatible, and aesthetic make them quite suitable for medical usage. Zirconia ceramic has bioinertness and noncytotoxicity. Carbon is another alternative with similar mechanical properties to bone, and it also features blood compatibility, no tissue reaction, and non-toxicity to cells. None of the three bioinert ceramics exhibit bonding with the bone. However, bioactivity of bioinert ceramics can be achieved by forming composites with bioactive ceramics. Bioglass and glass ceramics are nontoxic and chemically bond to bone. Glass ceramics elicit osteoinductive properties, while calcium phosphate ceramics also exhibit non-toxicity to tissues and bioresorption. The ceramic particulate reinforcement has led to the choice of more materials for implant applications that include ceramic/ceramic, ceramic/polymer, and ceramic/metal composites. Among these composites ceramic/polymer composites have been found to release toxic elements into the surrounding tissues. Metals face corrosion related problems, and ceramic coatings on metallic implants degrade over time during lengthy applications. Ceramic/ceramic composites enjoy superiority due to similarity to bone minerals, exhibiting biocompatibility and a readiness to be shaped. The biological activity of bioceramics has to be considered under various *in vitro* and *in vivo* studies. Performance needs must be considered in accordance with the particular site of implantation.

Processing

Technically, ceramics are composed of raw materials such as powders and natural or synthetic chemical additives, favoring either compaction (hot, cold or isostatic), setting (hydraulic or chemical), or accelerating sintering processes. According to the formulation and shaping process used, bioceramics can vary in density and porosity as cements, ceramic depositions, or ceramic composites.

A developing material processing technique based on the biomimetic processes aims to imitate natural and biological processes and offer the possibility of making bioceramics at ambient temperature rather than through conventional or hydrothermal processes [GRO 96]. The prospect of using these relatively low processing temperatures opens up possibilities for mineral organic combinations with improved biological properties through the addition of proteins and biologically active molecules (growth factors, antibiotics, anti-tumor agents, etc.). However, these materials have poor mechanical properties which can be improved, partially, by combining them with bonding proteins.

Commercial usage

Common bioactive materials available commercially for clinical use include 45S5 bioactive glass, A/W bioactive glass ceramic, dense synthetic HA, and bioactive composites such as a polyethylene–HA mixture. All these materials form an interfacial bond with adjacent tissue.

High-purity alumina bioceramics are currently commercially available from various producers. U.K. manufacturer Morgan Advanced Ceramics (MAC) began manufacturing orthopaedic devices in 1985 and quickly became a recognised supplier of ceramic femoral heads for hip replacements. MAC Bioceramics has the longest clinical history for alumina ceramic materials, manufacturing HIP Vitox® alumina since 1985. Some calcium-deficient phosphates with an apatite structure were thus commercialised as "tricalcium phosphate" even though they did not exhibit the expected crystalline structure of tricalcium phosphate.

Currently, numerous commercial products described as HA are available in various physical forms (e.g. granules, specially designed blocks for specific applications). HA/polymer composite (HA/polyethyelene, HAPEXTM) is also commercially available for ear implants, abrasives, and plasma-sprayed coating for orthopedic and dental implants.

CERAMIC NANOPARTICLE

Nanoceramic is a type of nanoparticle that is composed of ceramics, which are generally classified as inorganic, heat-resistant, nonmetallic solids made of both metallic and nonmetallic compounds. The material offers unique properties. Macroscale ceramics are brittle and rigid and break upon impact. However, nanoceramics take on a larger variety of functions, including dielectric, ferroelectric, piezoelectric, pyroelectric, ferromagnetic, magnetoresistive, superconductive and electro-optical.

Nanoceramics were discovered in the early 1980s. They were formed using a process called sol-gel which mixes nanoparticles within a solution and gel to form the nanoparticle. Later methods involved sintering (pressure and heat). The material is so small that it has basically no flaws. Larger scale materials have flaws that render them brittle.

In 2014 researchers announced a lasering process involving polymers and ceramic particles to form a nanotruss. This structure was able to recover its original form after repeated crushing.

Properties

Nanoceramics have unique properties because of their size and molecular structure. These properties are often shown in terms of various electrical and magnetic physics phenomenons which include:

- Dielectric: An electrical insulator that can be polarized (having electrons aligned so that there is a negative and positive side of the compound) by an electric field to shorten the distance of electron transfer in an electric current.

- Ferroelectric: Dielectric materials that polarize in more than one direction (the negative and positive sides can be flipped via an electric field).

- Piezoelectric: Materials that accumulate an electrical charge under mechanical stress.

- Pyroelectric: Material that can produce a temporary voltage given a temperature change.

- Ferromagnetic: Materials that can to sustain a magnetic field after magnetization.

- Magnetoresistive: Materials that change electrical resistance under an external magnetic field.

- Superconductive: Materials that exhibit zero electric resistance when cooled to a critical temperature.

- Electro-optical: Materials that change optical properties under an electric field.

Nanotruss

Nanoceramic is more than 85% air and is very light, strong, flexible and durable. The fractal nanotruss is a nanostructure architecture made of alumina, or aluminum oxide. Its maximum compression is about 1 micron from a thickness of 50 nanometers. After its compression, it can revert to its original shape without any structural damage.

Synthesis

Sol-gel

One process for making nanoceramics varies is the sol-gel process, also known as chemical solution deposition. This involves a chemical solution, or the sol, made of nanoparticles in liquid phase and a precursor, usually a gel or polymer, made of molecules immersed in a solvent. The sol and gel are mixed to produce an oxide material which are generally a type of ceramic. The excess products (a liquid solvent) are evaporated. The particles desires are then heated in a process called densification to produce a solid product. This method could also be applied to produce a nanocomposite by heating the gel on a thin film to form a nanoceramic layer on top of the film.

Two-photon Lithography

This process uses a laser technique called two-photon lithography to etch out a polymer into a three-dimensional structure. The laser hardens the spots that it touches and leaves the rest unhardened. The unhardened material is then dissolved to produce a "shell". The shell is then coated with ceramic, metals, metallic glass, etc. In the finished state, the nanotruss of ceramic can be flattened and revert to its original state.

Sintering

In another approach sintering was used to consolidate nanoceramic powders using high temperatures. This resulted in a rough material that damages the properties of ceramics and requires more time to obtain an end product. This technique also limits the possible final geometries. Microwave sintering was developed to overcome such problems. Radiation is produced from a magnetron, which produces electromagnetic waves to vibrate and heat the powder. This method allows for heat to be instantly transferred across the entire volume of material instead of from the outside in.

The nanopowder is placed in an insulation box composed of low insulation boards to allow the microwaves to pass through it.The box increases temperature to aid absorption. Inside the boxes are suspectors that absorb microwaves at room temperature to initialize the sintering process. The microwave heats the suspectors to about 600 °C, sufficient to trigger the nanoceramics to absorb the microwaves.

Applications

Medical technology used nanoceramics for bone repair. It has been suggested for areas including energy supply and storage, communication, transportation systems, construction and medical technology. Their electrical properties may allow energy to be transferred efficiencies approaching 100%. Nanotrusses may be eventually applicable for building materials, replacing concrete or steel.

MACHINING OF CERAMICS

Ceramics have a unique combination of mechanical, physical and chemical properties:

- High strength,

- Hardness,

- Low density,

- High stiffness (modulus of elasticity),

- Good tribological properties (e.g., excellent resistance to different types of wear},

- Very low electric conductivity,

- Very low thermal conductivity,

- High refractoriness and thermal stability,

- Good corrosion resistance.

The techniques of manufacturing ceramics (shape forming, sintering) allow to produce the parts of different sizes and shapes with good cost efficiency.

However the machining of ceramics is very expensive and time consuming operation representing from 50 to 90% of the total cost of the part.

The main parameter of the efficiency of a machining operation is the Material Removal Rate (MRR) indicating the volume of the material removed from the workpiece surface for 1 min.

The methods of ceramics machining:

- Machining of ceramics in the presintered state,

- Grinding of ceramics,

- Ultrasonic machining of ceramics,

- Rotary ultrasonic machining of ceramics,

- Laser assisted machining of ceramics,

- Laser machining of ceramics.

Machining of Ceramics in the Presintered State

Sintered ceramics are very hard and therefore their machining is an expensive, difficult and time consuming process.

Ceramic parts may be effectively machined before the final sintering stage either in the "green" (non-sintered powder) compact state or in the presintered "bisque" state.

Conventional machining methods (milling, drilling, turning) may be applied for the ceramic parts in the presintered state.

Titanium nitride (TiN) coated high speed steel tools, tungsten carbide tools and poly-crystalline diamond (PCD) tools are used in machining of presintered ceramics.

The material removal rate (MRR), which may be achieved in machining of ceramics in the bisque (presintered) state is 0.6 in^3/min (9832 mm^3/min). This value is similar or even higher than MRR of Tool and die steels.

Shrinkage and warping of the ceramic part occurring during its sintering does not allow to achieve tight dimensional tolerances and high quality surface finish in the presin-tered state.

However machining of bisque ceramics allows to reduce the cost of the final machining of the parts in the sintered state.

Grinding of Ceramics

Grinding is the most widely used method of machining of Ceramics in the sintered state.

Grinding operation involves a rotating abrasive wheel removing the material from the surface of the workpiece. The grinding zone is continuously flushed with a fluid coolant, which cools the grinding zone, lubricates the contact between the wheel and the part surfaces, removes the micro-chips (debris) produced in the grinding process.

Resin-bond wheels with either synthetic or natural diamond of different grit size pressed at different concentrations in polymer (resin) matrices are commonly used for grinding ceramics.

Electrolytic in-process dressing (ELID) technique of dressing metal-bonded grinding wheels is used for fine (nano) finish grinding.

The Material Removal Rate (MRR) of grinding ceramics is maximum 0.0006 in^3/min (9.832 mm^3/min).

Ultrasonic Machining of Ceramics

Ultrasonic machining (UM) of ceramics is the machining method using the action of a slurry containing abrasive particles flowing between the workpiece and a tool vibrating at an ultrasonic frequency.

The vibration frequency is 19 ~ 25 kHz.

The amplitude of vibration 0.0005 - 0.002" (13 – 50 μm). During the operation the tool is pressed to the workpiece at a constant load. The slurry strike the ceramic workpiece and remove small ceramic debris fracturing from the surface.

Conventional ultrasonic machining (UM) is characterized by low material removal rates: up to 0.003 in³/min (49 mm³/min).

Other disadvantages of the conventional ultrasonic machining method are low accuracy and high tool wear.

Ultrasonic machining is used commonly for drilling operation.

Rotary Ultrasonic Machining of Ceramics

Rotary ultrasonic machining (RUM) of ceramics combines grinding operation with the method of ultrasonic machining.

A core drill tool made of a metal bonded diamond grits is used in the rotary ultrasonic machining (commonly drilling).

The tool is rotating and simultaneously vibrating at an ultrasonic frequency. The tool is continuously fed and pressed at a load towards the ceramic workpiece causing abrasive action performed by the rotating-vibrating diamond grits.

A fluid coolant is continuously flowing through the core of the tool to the grinding zone cooling it and removing the debris produced in the grinding process.

Rotary ultrasonic machining is much more effective than conventional ultrasonic machining.

The RUM material removal rate is up to 0.03 in³/min (492 mm³/min).

Laser Assisted Machining of Ceramics

Laser assisted machining (LAM) is the method of machining ceramics using a laser beam directed to the workpiece area located directly in front of the conventional cutting tool.

The laser beam heats and softens (not melts) the ceramic material at the surface just prior the cutting action.

As a result the cut material becomes ductile and it may be removed much faster than in conventional cutting operation without a laser assistance.

The LAM material removal rate is up to 0.06 in³/min (983 mm³/min).

Titanium nitride coated tools are used for the laser assisted machining of ceramics.

Traditional machining operations (milling, turning) may be performed by the method of the laser assisted machining.

Laser Machining of Ceramics

Laser machining of ceramics is the machining operation performed by a high power laser melting the material, which is blown away by a supersonic gas jet.

The laser energy density required for melting alumina ranges from 4.6 BTU/in² to 6.1 BTU/in² (750 J/cm² to 1000 J/cm²).

The following machining operation may be performed by laser:

- Drilling,

- Cutting,

- Scribing and marking.

Residual stresses and micro-cracks may form at the cut edge as a result of the shrinkage of the solidified molten material.

Preheating of the ceramic workpiece to 2550 °F (1399 °C) prior to the laser machining allows to minimize micro-cracking due to reduction of the temperature gradients and thermal stresses.

IONIC AND COVALENT BONDING

Ceramics (ceramic materials) are non-metallic inorganic compounds formed from metallic (Al, Mg, Na, Ti, W) or semi-metallic (Si, B) and non-metallic (O, N, C) elements.

Atoms of the elements are held together in a ceramic structure by one of the following bonding mechanism: Ionic Bonding, Covalent Bonding, Mixed Bonding (Ionic-Covalent).

Most of ceramic materials have a mixed bonding structure with various ratios between Ionic and Covalent components. This ratio is dependent on the difference in the electronegativities of the elements and determines which of the bonding mechanisms is dominating ionic or covalent.

- Electronegativity,

- Ionic Bonding,

- Covalent Bonding,

- Ionic-Covalent (mixed) Bonding,

- Characterization of ceramics properties.

Electronegativity

Electronegativity is an ability of atoms of the element to attract electrons of atoms of another element. Electronegativity is measured in a relative dimensionless unit (Pauling scale) varying in a range between 0.7 (francium) to 3.98 (fluorine).

Non-metallic elements are strongly electronegative. Metallic elements are characterized by low electronegativity or high electropositivity – ability of the element to lose electrons.

Ionic Bonding

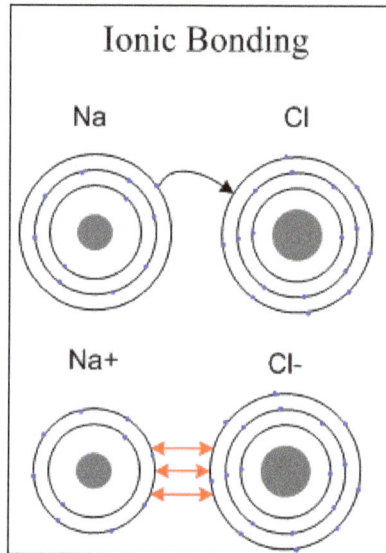

Ionic bonding occurs between two elements with a large difference in their electronegativities (metallic and non-metallic), which become ions (negative and positive) as a result of transfer of the valence electron from the element with low electronegativity to the elem.1ent with high electronegativity.

The typical example of a material with Ionic Bonding is sodium chloride (NaCl).

Electropositive sodium atom donates its valence electron to the electronegative chlorine atom, completing its outer electron level (eight electrons).

As a result of the electron transfer the sodium atom becomes a positively charged ion (cation) and the chlorine atom becomes a negatively charged ion (anion). The two ions attract to each other by Coulomb force, forming a compound (sodium chloride) with ionic bonding. Ionic bonding is non-directional.

Covalent Bonding

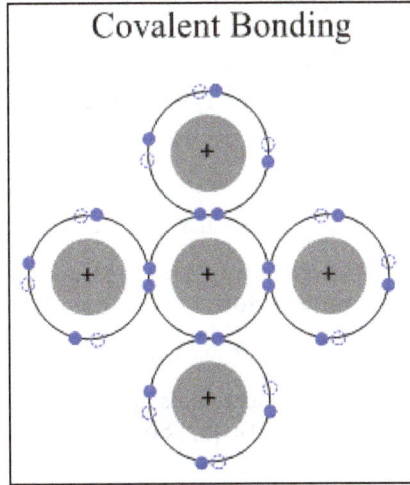

Covalent bonding occurs between two elements with low difference in their electronegativities (usually non-metallics), outer electrons of which are shared between the four neighboring atoms.

Covalent Bonding is strongly directional.

Ionic-covalent (Mixed) Bonding

Ionic-covalent (mixed) bonding with various ratios of the two fractions (ionic and covalent) occurs in most of ceramic materials.

Degree of Ionic Bonding can be estimated from the following formula:

I.F. = $\exp(-0.25 * \Delta E^2)$

Where:

I.F. – fraction of ionic bonding,

ΔE – difference in the electronegativities of the elements.

Characterization of Ceramics Properties

In contrast to metallic bonding neither ionic nor covalent bonding form free electrons, therefore ceramic materials have very low electric conductivity and thermal conductivity.

Since both ionic and covalent bonds are stronger than metallic bond, ceramic materials are stronger and harder than metals.

Strength of ionic and covalent bonds also determines high melting point, modulus of elasticity (rigidity), temperature and chemical stability of ceramic materials.

Motion of dislocations through a ceramic structure is impeded therefore ceramics are generally brittle that limits their use as structural materials.

Ceramics may have either crystalline or amorphous structure. There are also ceramic materials, consisting of two constituents: Crystalline and amorphous.

References

- Ceramic-coating-metals, chemicals: thomasnet.com, Retrieved 2 August, 2019

- Boch, Philippe, Niepce, Jean-Claude. (2010) Ceramic Materials: Processes, Properties and Applications. Doi: 10.1002/9780470612415.ch12

- Fracture-toughness-tests-of-ceramics: substech.com, Retrieved 30 March, 2019

- Chai, Chou; Leong, Kam W (2007). "Biomaterials Approach to Expand and Direct Differentiation of Stem Cells". Molecular Therapy. 15 (3): 467–80. Doi:10.1038/sj.mt.6300084. PMC 2365728. PMID 17264853

- Ceramic-engineering, entry: newworldencyclopedia.org, Retrieved 29 August, 2019

- Claire Diop, Julie. "R&D 2002:Nano Ceramics". MIT Technology Review. Retrieved December 1, 2002

- Machining-of-ceramics, dokuwiki: substech.com, Retrieved 6 January, 2019

- Ionic-and-covalent-bonding, dokuwiki: substech.com, Retrieved 11 June, 2019

5

Applications of Ceramics

Ceramics are applied in a wide range of fields such as electronics, dentistry, art, transportation, high pressure fuel systems, thermal insulators, chemical industry, pharmaceutical industry and automotive engineering. The diverse applications of ceramics in these fields have been thoroughly discussed in this chapter.

CERAMICS IN ELECTRONICS

Products made of technical ceramics are now proven components in the construction and control of sophisticated plants, machinery and equipment with electrotechnical component assemblies. Often, they make possible the function of such constructions in the first place. Typical examples include λ sensors in automotive engineering or in kiln and furnace engineering, the vacuum chambers of particle accelerators or actuators in motion detectors. The size of such components is typically in the region of a few millimetres up to several metres.

A special feature of this class of materials is the wide range of electrical conductivity, which spans more than 15 orders of magnitude and cannot be matched by any other class of materials. It includes electrically insulating as well as semi-conducting, ionic-conducting and superconducting materials. On top of this come the dielectric properties, which can be used, for instance, in sensor technology and telecommunications.

Besides the electrical properties, magnetic properties are often required, The soft or hard magnetic ferrites have proven effective materials for decades. Compared with metallic materials, they often enable smaller product sizes and therefore more economically attractive products.

Typical for the applications of technical ceramic materials is a frequent requirement for other non-electrical properties such as:

- Mechanical strength,
- Thermal resistance,
- Thermal shock resistance,

- Thermal conductivity,

- Corrosion resistance,

- Production of ultrahigh-vacuum-capable joints with metals.

A key strength of this class of materials is the demand-driven combinations of the above-mentioned properties. In addition comes the possibility to optimize properties for a specific application by means of appropriate doping and therefore to tailor materials to requirements.

For instance, the zirconia ceramics used for λ sensors can be optimized to maximize their strength without compromising their suitability for use as oxygen sensors as only one electrical signal must be recorded, which can be evaluated on the basis of appropriate calibration. With the selective modification of the chemical composition of this material, maximized electrical conductivity can be obtained along with an acceptable level of strength. As a result, this anionic conductor can be efficiently used as electrolyte for SOFCs (high-temperature fuel cells).

Doping of the electrically insulating Al_2O_3 with sodium forms a superionic conductor for Na^+, which can be used as a separator in Na/S high-energy batteries.

Al_2O_3 ceramic with purity up to > 99,9 % remains one of the most commonly used oxide ceramic electrical insulating materials even today. Its specific electrical resistance reaches 10^{16} Ω*cm at room temperature, while at 1600 °C it still reaches a level of 10^6 Ω*cm. For this reason, these materials have been proven components for electrical insulation especially in high temperatures for decades.

The electrical properties of technical ceramics in combination with their mechanical properties generally enable a reduction in the sizes of existing systems, often simplified designs and therefore significantly enhanced performance. Typical examples are electrical feedthroughs on Al_2O_3 basis to withstand pressure above 1000 bar or substrates made of AlN ceramic which thanks to their high thermal conductivity of more than 200 W*(m*K)$^{-1}$ are used in demanding electronic applications.

The wide-ranging application of technical ceramic materials in electrical engineering and electronics has led over the past decades to a correspondingly high number of variants in electrically passive and active materials. Today a broad spectrum exists including insulators, dielectrics, piezoelectrics, NTC and PTC ceramics, varistors, cationic, anionic and electron conductors on oxide and non-oxide basis, superconductors as well as soft and hard magnetic ferrites.

The use of ceramic components in electrotechnical assemblies often demands material-to-material joints with one or more metallic components that can comprise different materials. Such joints are usually associated with the demand for high or ultrahigh vacuum tightness and acceptable mechanical strength for use in the field. Such ceramic metal joints are generally realized by hard brazing based on MoMn procedures. For this purpose a thick film of molybdenum applied and then fired onto the ceramic is joined with the respective metal part with a silver-based solder as standard or, if corrosive attack is expected, with gold-based solder. This process can be generally applied for all types of Al_2O_3 ceramics, but only certain types of ZrO_2 ceramics. Several decades ago, an active soldering process was developed that enables direct soldering of metal components with ceramics without prior metallization. This process is applied today as an alternative to metallized ceramics, mainly for ZrO_2. With both processes, given appropriate design and selection of the ceramic and metallic materials, joint strengths exceeding 200 MPa are achieved.

The nearly $4.5 trillion global electronics industry would not exist without ceramics. Ceramic-based components are indispensable in products such as smartphones, computers, televisions, automotive electronics, and medical devices.

Although ceramics have traditionally been considered insulating materials, after World War II, research in material science has led to the development of new ceramic formulations that exhibit semiconducting, superconducting, piezoelectric, and magnetic properties.

Ceramic products used as electrical insulators include spark plugs, hermetic packaging, ceramic arc tubes, and protective parts (e.g., beads and tubing) for bare wires and power lines. These products are primarily used in sectors such as automotive, marine transportation, aerospace, and electricity distribution. Among these products, spark plugs represent the oldest and the most popular. They were first invented in 1860 to ignite fuel for internal combustion engines and are still being used for this purpose today. However, as the automotive industry shifts toward electric and hybrid vehicles, demand for ceramic spark plugs is expected to drop. The main suppliers of these products are directing their know-how in electroceramics toward other components such as ceramic solid-state batteries.

From a market standpoint, passive components are experiencing the most demand. Passive components are referred to as such because they are not able to control the flow of electrons in a circuit, unlike active components (also known as semiconducting devices). Passive components are only used to store or transform energy. Ceramic capacitors represent the bulk of the passive component market and have experienced rapid growth lately due to strong demand from the mobile device and communications sector. These capacitors utilize the insulating properties of a ceramic material (called dielectric) placed between two or more metal layers to store electrical charges. Ceramic capacitors are becoming very small in size, almost microscopic. Recently, multilayer ceramic capacitors (MLCCs) that measure only 0.25mm by 0.125 mm by 0.125 mm have been commercialized.

Other types of passive components include fixed and variable resistors (they use ceramics to dissipate energy), inductors (they use a magnetically permeable ceramic to create a magnetic field), and circuit protection devices (they use a semiconducting ceramic to protect the circuit against transient voltage surges).

Piezoelectric ceramics are materials that generate an electrical charge when pressure is applied or change size under an electric field. They are used to transform mechanical energy into electric energy or vice versa. These devices are becoming increasingly popular for use as filters, resonators, transducers, acoustic elements, actuators, and components for pressure sensors. Easy to manufacture in various shapes and sizes, piezoelectric ceramics contribute significantly to the miniaturization of electro-mechanical features and, consequently, are gaining greater penetration in the fields of consumer electronics, robotics, automotive, sensors and instrumentation, and energy harvesting.

Other types of electroceramics consist primarily of ferrite-based permanent magnets and circuit devices for high-reliability applications (e.g., low- and high-temperature co-fired ceramics, and ceramic electronic substrates).

Within the electronics sector, glass is primarily used to fabricate display panels for televisions, computers, and mobile devices. The unprecedented mass appeal of portable devices has led producers to create ultra-tough glass that does not break when dropped.

Glass powder is also added in thick film paste for metallization of electronic components, whereas glass seals are used in certain applications to protect electronic devices from the environment. Flexible glass is being produced for fabrication of flexible devices, such as organic light emitting diodes (OLEDs).

The main applications of ceramics and glass in electrical and electronic applications are illustrated below.

Main applications of ceramics and glass in electrical and electronic applications:

Ceramics

- Transportation electrical systems:
  - Spark plugs.
- Power distribution:
  - Protective parts for power lines.
- Electronic circuits:
  - Capacitors,
  - Resistors,
  - Inductors,

- ○ Circuit protection devices (for example, metal oxide varistors),

- ○ High and low-temperature co-fired ceramics.

- Electromechanical applications:

  - ○ Piezoelectric devices.

- Audio systems:

  - ○ Magnets.

Glass

- Displays:

  - ○ Rigid panels,

  - ○ Flexible panels.

CERAMIC KNIFE

A ceramic knife is a knife designed with a ceramic blade typically made from zirconium dioxide (ZrO_2; also known as zirconia). These knife blades are usually produced through the dry-pressing and firing of powdered zirconia using solid-state sintering. It is 8.5 on the Mohs scale of mineral hardness, compared to 4.5 for normal steel and 7.5 to 8 for hardened steel and 10 for diamond. The resultant blade has a hard edge that stays sharper for longer when compared to conventional steel knives. Whilst the edge is harder than a steel knife, it is less tough and thus more brittle. The ceramic blade is sharpened by grinding the edges with a diamond-dust-coated grinding wheel.

Zirconium Oxide

A ceramic knife made from blackened zirconia,
super heated under pressure.

Zirconium oxide is used due to its polymorphism. It exists in three phases: monoclinic, tetragonal and cubic. Cooling to the monoclinic phase after sintering causes a large volume change, which often causes stress fractures in pure zirconia. Additives such as magnesia, calcia and yttria are used in the manufacture of the knife material to stabilize the high-temperature phases and minimize this volume change. The highest strength and toughness is produced by the addition of 3 mol% yttrium oxide yielding partially stabilized zirconia. This material consists of a mixture of tetragonal and cubic phases with a bending strength of nearly 1,200 MPa. Small cracks allow phase transformations to occur, which essentially close the cracks and prevent catastrophic failure, resulting in a relatively tough ceramic material, sometimes known as TTZ (transformation-toughened zirconia).

Properties

Ceramic knives are substantially harder than steel knives, will not corrode in harsh environments, are non-magnetic, and do not conduct electricity at room temperature. Because of their resistance to strong acid and caustic substances, and their ability to retain a cutting edge longer than forged metal knives, ceramic knives are better suited for slicing boneless meat, vegetables, fruit and bread. Since ceramics are brittle, blades may break if dropped on a hard surface although improved manufacturing processes have reduced this risk. They are also unsuitable for chopping through bones, or frozen foods, or in other applications which require prying, which may result in chipping. Several brands now offer either a black-coloured or a designed blade made through an additional hot isostatic pressing step, which increases the toughness.

Sharpening and General Care

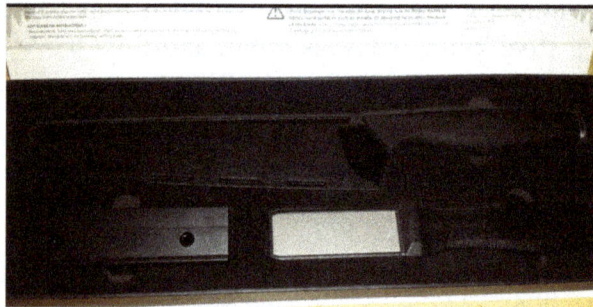

A diamond dust sharpener with cradle for a ceramic knife.

Unlike a traditional steel blade that benefits from regular honing and resharpening in order to keep a sharp edge, a ceramic knife will stay sharp and retain its cutting edge for much longer—up to 10x longer according to some tests. The inherent hardness of the ceramic material also makes it more difficult for the consumer to resharpen. Although a ceramic knife therefore does not need regular sharpening in the same way as steel, its blade edge will eventually degrade or chip and lose its cutting edge, at which point specialized sharpening services are required for the ceramic edge.

CERAMICS IN DENTISTRY

It is quite usual in dentistry to adopt a material from engineers and adapt it to clinical conditions. A good example of such an instance is dental ceramics. In Dental science, ceramics are referred to as nonmetallic, inorganic structures primarily containing compounds of oxygen with one or more metallic or semi-metallic elements. They are usually sodium, potassium, calcium, magnesium, aluminum, silicon, phosphorus, zirconium & titanium.

As we peep into the dental history, a French dentist De Chemant patented the first porcelain tooth material in 1789. In 1808 Fonzi, an Italian dentist invented a "terrometallic" porcelain tooth that was held in place by a platinum pin or frame. Ash developed an improved version of the platinum tooth in 1837. Dr. Charles Land patented the first Ceramic crowns in 1903.Vita Zahnfabrik introduced the first commercial porcelain in 1963.

Structurally, dental ceramics contain a crystal phase and a glass phase based on the silica structure, characterized by a silica tetrahedra, containing central Si^{4+} ion with four O-ions. It is not closely packed, having both covalent and ionic characteristics. The usual dental ceramic, is glassy in nature, with short range crystallinity. The only true crystalline ceramic used at present in restorative dentistry is Alumina (Al_2O_3), which is one of the hardest and strongest oxides known. Ceramics composed of single element are rare. Diamond is a major ceramic of this type, hardest natural material used to cut tooth enamel. Ceramics are widely used in dentistry due to its dual role – strength and esthetics.

Basically the inorganic composition of teeth and bones are ceramics – Hydroxyapatite. Hence ceramics like hydroxyapatite, wollastonite etc are used as bone graft materials. They have an entire plethora of synthetic techniques like wet chemical, sol-gel, hydrothermal methods etc. Also they are added as bioactive filler particles to other inert materials like polymers or coated over metallic implants. These ceramics are collectively called as bioceramics. There are basically two kinds of bioceramics-inert (e.g. Alumina) and bioactive (hydroxyapatite). They can be resorbable (Tricalciumphosphate) or non-resorbable (Zirconia).

Dental cements are basically glasses. Initially, silicate cements were introduced. They constitute the first dental cement to use glass as its component. The cement powder contains a glass of silica, alumina and fluorides. The liquid, is an aqueous solution of phosphoric acid with buffer salts. Fluoride ions leached out from the set cements are responsible for the anticariogenic property. But silicates are discontinued due to low pH during setting reaction that affects the dental pulp.

In cements called glass ionomers, the glass forms the filler and acidic polymers form the matrix. The reaction that proceeds is usually an acid base reaction. Usually they are dispensed as powder containing glass powder, and liquid containing an acid, say, polyacrylic acid. When they are mixed, the acid in liquid etches the glass and reacts with calcium and other ions forming salts and the cement sets to a hard mass. Cements are direct restorative materials – i.e. manipulated and placed onto teeth directly, unlike other restorations that are made outside and fixed to teeth, called indirect restorations. They are highly advantageous as they are quick to set, release fluoride leading to anti-cariogenic action, esthetic and chemically bond to tooth material.

The use of ceramics are encouraged by their biocompatibility, aesthetics, durability and easier customization. The specialty of ceramic teeth is the ability to mimic the natural tooth in colour and translucency along with strength. Ceramics have excellent intraoral stability and wear resistance adding to their durability.

Dental ceramics, since introduction have undergone numerous modifications in terms of chemistry. Ceramics have been able to give heed to the ever changing needs in dentistry. To delve deep into the relevance of ceramic in dentistry, one should understand the physics of forces acting in the oral cavity. The masticatory (chewing) force is the strongest force present here. Other minor forces include that of tongue and periodontal ligament, which do not relate to the use of ceramics in dentistry.

The masticatory force is generated outside oral cavity by basically strong muscles, that move the jaw, open it or close it. Closure of jaw produces two kinds of forces. It is predominantly compressive in nature. Frequently impact kind of force is also experienced. Hence a ceramic has to undergo cycles of these forces indefinitely, without fracture, to result in a successful restoration of lost teeth structures.

In order to have a complete idea of what ceramic means to dentistry, we need to look at the complete range of ceramics used in this discipline.

Classification of Dental Ceramics

Classification of ceramics in dentistry is apparently an impossible task due to vast improvements made in the compositions. Nevertheless, the table provided here gives a general idea, say a bird's eye view of ceramics in dentistry.

Microstructural Classification

- Category 1: Glass-based systems (mainly silica).
- Category 2: Glass-based systems (mainly silica) with fillers usually crystalline (typically leucite or a different high-fusing glass).
  - Low-to-moderate leucite-containing feldspathic glass.
  - High-leucite (approx. 50%)-containing glass, glass-ceramics (Eg: IPS Empress).
  - Lithium disilicate glass-ceramics (IPS e.max® pressable and machinable ceramics).

- Category 3: Crystalline-based systems with glass fillers (mainly alumina).

- Category 4: Polycrystalline solids (alumina and zirconia).

Based on Processing Technique

- Powder/liquid glass-based systems.

- Pressable blocks of glass-based systems.

- CAD/CAM systems.

Based on Composition

- Silicates: These are characterized by amorphous glass phase, containing predominantly silica.

- Oxide ceramics: It is notable that only oxide ceramics are used in dentistry, since nonoxide ceramics are difficult to process. Oxide ceramics contain a principal crystalline phase like Alumina.

- Zirconia has very high fracture toughness.

- Glass ceramics: These are type of ceramics that contains a glass matrix phase & at least one crystal phase.

- Although classification of dental ceramics based on composition is not much of importance today, due to advances made, it is included for historic importance.

Based on Type

- Feldspathic porcelain. Leucite – reinforced porcelain, Aluminous porcelain.

- Glass infiltrated alumina, Glass infiltrated zirconia. Glass ceramics.

Based on Firing Temperature

- Ultra-low fusing < 850°C.

- Low fusing 850°C - 1100°C.

- Medium Fusing 1101°C - 1300°C.

- High fusing >1300°C.

Based on Substructure Metal

- Cast Metal, Swaged metal, Glass ceramics. Sintered core ceramics and CAD-CAM porcelain. The various types of metals in metal ceramics include noble alloys like gold alloys, base metals like iron, indium & tin. Pure metals like commercially pure titanium, platinum, gold and palladium alloys and Base metal alloys (nickel, chromium).

Based on Reinforcing Method

- Reinforced ceramic core systems.

- Resin-bonded ceramics.

- Metal–ceramics.

Table: Typical oxide composition of dental porcelain.

Material	wt %
Silica	~ 62
Alumina	~ 18
Boric oxid	~ 7
Potash (K_2O)	~ 7
Soda (Na_2O)	~ 4
Other oxides	~ 2

The classification based on microstructure will be dealt with in detail and an idea of classification by processing technique is added in this review. Other classifications are of academic interest.

Basically ceramics are used as indirect restorative materials such as crowns and bridges, Inlays/Onlays and dental implants. Recently ceramic braces are used in orthodontics.

Crowns and Bridges

Crown is technically a "Cap" placed on a tooth to protect it from fracture or sensitivity. On the other hand a bridge is a fixed replacement of missing teeth, with support from adjacent teeth. Both of these function similarly on biological and biomechanical terms, hence discussed together. These are of either Porcelain fused to metal (PFMs) or full ceramics. In case of PFMs, a metal core is placed in the tooth surface and ceramic is built on it. This is done by initially preparing the metallic portion by conventional casting techniques. Then the ceramic powder is incrementally painted on it and sintered. This is followed by glazing. In case of full ceramics, the wax pattern is prepared for the crown, it is invested and mould space prepared by lost wax technique. The ceramic is fused and typically pressed into the mould space.

Similarly, ceramic teeth are manufactured in various shades, shapes and sizes to be used in complete dentures. Also, in case of gum recession, in fixed prosthesis, pink coloured ceramics are placed in the lost gum region to make it look natural.

Category 1:Glass-based Systems (Mainly Silica):

Chemical composition of these ceramics is based on silica network and potash feldspar ($K_2O.Al_2O_3.6SiO_2$) or soda feldspar ($Na_2O.Al_2O_3.6SiO_2$) or both. Potassium and sodium feldspar are naturally occurring minerals composed primarily of Potash and soda. The most important property of feldspar is its tendency to form crystalline mineral leucite when melted.

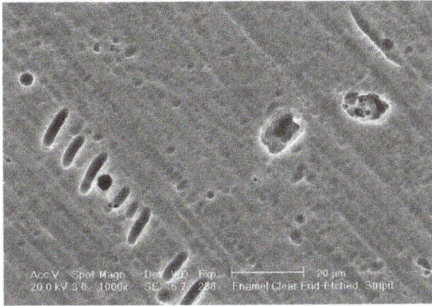

Scanning electron micrograph
of amorphous glass.

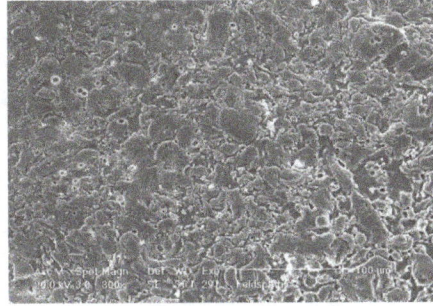

SEM image of Feldspathic
porcelain.

Mechanical properties say, flexural strength usually ranges from 60 MPa to 70 MPa. Hence, they can be used as veneer materials for metal or ceramic substructures, as well as for veneers, using either a refractory die technique or platinum foil.

Then, few other components like pigments, opacifiers and glasses are added to control fusion and sintering temperature, thermal properties and solubility. Glass modifiers like boric oxide can be added to reduce viscosity and softening temperature. Pigments are basically metallic oxides – say nickel oxide, manganese oxide, etc. Tin oxide is used for opaquing, iron oxide used for brown, copper oxide for green, titanium oxide for yellow, manganese oxide for purple, cobalt oxide for blue, nickel oxide for brown and rare earth oxides for simulating ultraviolet reflectance of natural teeth in ceramic.

The pigment oxides are also called as colour frits. These are added in appropriate proportions, dictated by intensity of colour required. Then the material is fired and fused to form glass, that is powdered again.

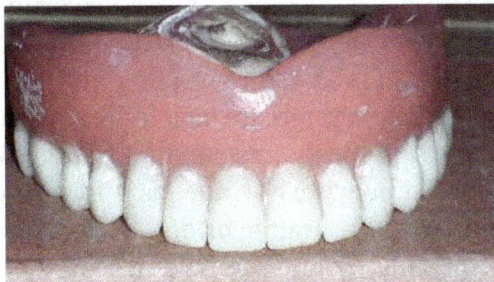

Acrylic Denture with Ceramic teeth
in lab processing stage.

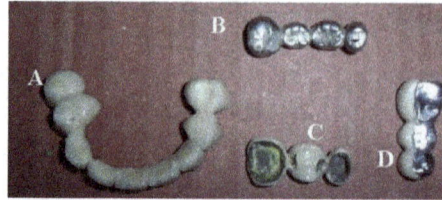

Different kinds of fixed Dentures A)PFM Bridge, B) Full metal bridge,
C)Inner side of PFM bridge, D) Metal bridge with ceramic facing.

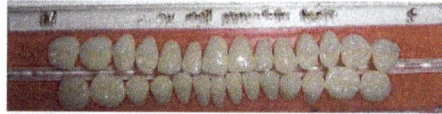

Porcelain teeth set for removable prosthesis.

These ceramics are strengthened by either development of residual compressive stresses within the surface or by interruption of crack propagation through the material. Residual compressive stresses are introduced by ion exchange and thermal tempering. Interruption of crack propagation is in turn achieved by dispersion of crystalline phases like partially stabilized Zirconia.

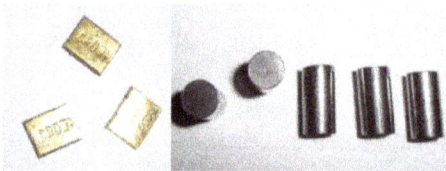

Representative alloys
for PFM cores.

Commonly used dental shade guide showing
corresponding numbers for shades.

Glass ceramics made of a glass matrix phase and at least one crystal phase is produced by controlled crystallization of glass. It is available as castable machinable, pressable and infiltrated forms used in all-ceramic restorations. The first commercially available castable glass ceramic was developed by the Corning Glass Works (Dicor) in 1950s. It paved way to dental ceramic system relying upon strengthening of glass with various forms of mica. ($SiO_2.K_2O.MgO.Al_2O_3.ZrO_2$, with the addition of some fluorides). Fluorides present in these ceramics are responsible for their nature-like fluorescence. Mould space is created by lost wax method and desired shape is formed. This process followed by coating with veneering porcelain. The noteworthy aspect of this ceramic is the Chameleon effect in which some part of color is picked up from adjacent teeth. Here, the ceraming process results in the nucleation and the growth of tetrasilicate mica crystals within the glass. The crystals are needle-like in shape and arrest the propagation of cracks. Mechanical property measurements suggest the flexural strength is in the range of 120–150MPa, may just be adequate for posterior crowns but is not sufficient for the construction of all-ceramic bridges. The passage of light through the material is affected by the crystal size and the difference in the refractive indices of the glass phase and the crystalline phase. If the crystals are smaller than the wavelength of visible light (0.4–0.7mm) the glass will appear transparent.

The refractive index of the small mica crystals is closely matched to that of the surrounding glass phase, such that the tendency for light to scatter is lower than the aluminous porcelains.

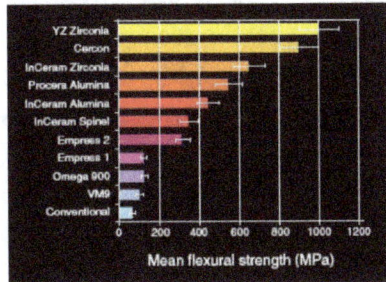

Flexural strengths of various ceramics.

The Machinable glass ceramic is another high quality product, which is crystallized during the manufacture and provided as CAD/CAM blanks or ingots. They provide better precision than castable glass ceramic, due to reduction of casting errors. They exhibit similar mechanical properties, to castable forms but are less translucent.

Category 2: Glass-based systems (mainly silica) with fillers usually crystalline (typically leucite or a different high-fusing glass):

This is a modification to category 1, in that varying amounts of other crystals are added or grown. The primary crystal types are either leucite, lithium disilicate, or fluorapatite. Leucite has been widely used as a constituent of dental ceramics to modify the coefficient of thermal expansion. This is most important where the ceramic is to be fused or baked onto metal (Optec HSP). But in leucite reinforced ceramic system, IPS Empress, leucite has a different role. This material relies on an increased volume of fine leucite particles to increase flexural strength. Leucite is nothing but potassium aluminum silicate mineral with large coefficient of thermal expansion compared with glasses. The property of Feldspar to form Leucite is exploited in the manufacture of porcelains for metal bonding. Newer generations of materials have much finer leucite crystals (10 μm to 20 μm) and even particle distribution throughout the glass. These materials are less abrasive and have much higher flexural strengths. In Figure, a scanning electron micrograph (SEM) of a typical feldspathic porcelain reveals a glass matrix surrounding leucite crystals. These materials are most commonly used as veneer porcelains for metal-ceramic restorations.

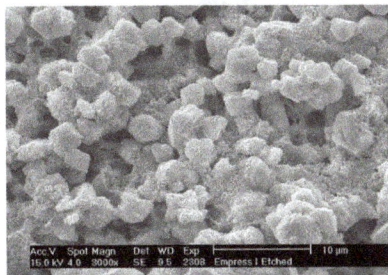

SEM Image of IPS-Empress I – Etched.

In the sintering process (Fortress and Optec-HSP), slurry of the ceramic powder is applied to a refractory die (unlike Platinum foil coated die in the procedure of the porcelain jacket crown), dried and subsequently fired in a porcelain furnace. Multiple layers can be built up to develop characterisation. Great skill is required by the dental laboratory technician to get the best aesthetics and appropriate contour. However, the strength of Leucite ceramics are insufficient for all-ceramic restorations.

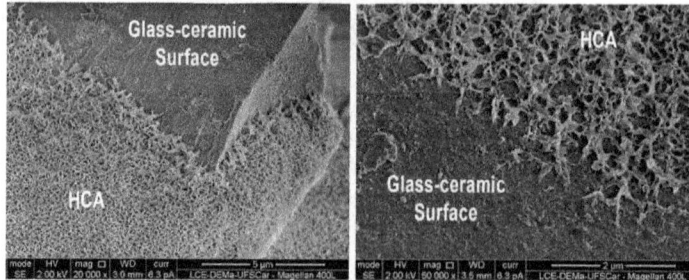

XRD of Optec HSP.

Lithium Disilicate and Apatite Glass Ceramics are based on SiO_2-Li_2O. The crystalline phase, lithium disilicate ($Li_2Si_2O_5$) makes up about 70% of the volume of the glass ceramic. Lithium disilicate has microstructure consisting of numerous small plate-like crystals that are interlocking and randomly oriented. This has a reinforcing effect on strength because, the needle-like crystals deflect cracks and arrest the propagation of cracks. A second crystalline phase, consisting of a lithium orthophosphate (Li_3PO_4) of a much lower volume, is present. The mechanical properties of this glass ceramic are far superior to that of the leucite glass ceramic, with a flexural strength in the region of 350– 450MPa and a fracture toughness approximately three times that of the leucite glass ceramic. There is a possibility for its use in all-ceramic systems. Also, processing is done by hot pressing technique.

For the alumina-based core systems feldspathic glasses can be used to provide the aesthetic surface layer, as their coefficients of expansion are closely matched. For the leucite glass ceramics the layering ceramic is identical to the core ceramic and so a mismatch in coefficient of expansion does not arise. However, this is not possible for lithium disilicate glass ceramic due to its higher coefficient of expansion. Here emerges a need for new compatible layering ceramic. This new layering ceramic is an apatite glass ceramic. The crystalline phase formed on ceraming is hydroxyapatite ($Ca_{10}(PO_4)_6OH_2$), which is the basic constituent of enamel. Thus, it represents a material that, at least in composition, is the closest match to enamel.

A veneer porcelain made of fluorapatite crystals in an aluminosilicate glass may be layered on the core to create the final morphology and shade of the restoration. The shape and volume of crystals increase the flexural strength to approximately 360 MPa, or about three times that of Empress. This material can be translucent even with the high crystalline content; this is due to the relatively low refractive index of the lithium-disilicate crystals. The material is translucent enough that it can be used for full-contour restorations or for the highest esthetics and can be veneered with special porcelain. Veneer porcelain consisting of fluorapitite crystals in an aluminosilicate glass may be layered on the core to create the final morphology and shade of the restoration. The fluorapatite (fluoride-containing calcium phosphate, $Ca_5(PO_4)_3F$) crystals contribute to the veneering porcelain's optical properties and CTE, so it matches the lithium-disilicate pressable or machinable material. Both the veneering and lithium-disilicate materials are etchable due to the glassy phase. Initial clinical data for single restorations are excellent with this material, especially if it is bonded. A material with similar properties and structure called 3G OPC is available as a pressable glass-ceramic from Pentron.

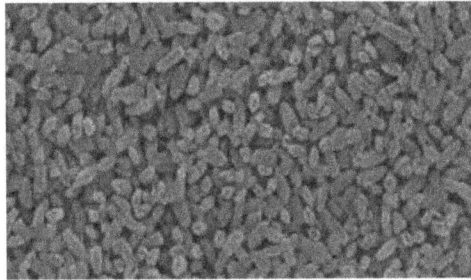

SEM image of IPS e.max.

Advantage of PFM is its high strength in clinical service. It uses the sub-structure metal to withstand stresses. Thermal compatibility is good. Lower crack propagation leading to better fracture resistance. But, on the other side of the coin, there is inadequate structure for ceramics due to thickness of metal. This results in reduction of more tooth structure while preparing it for the restoration. Metallic hue can be visible in anterior teeth. Metal will be exposed in case of gingival recession. Patients with allergy to metals can react adversely. As we add more layers between materials, more number of fractures can occur. Bonding failures at porcelain-metal interface can occur due to oxidation of interface metallic surface.

It is good to have a glance into preparation of tooth structure for a full crown, to understand the pros and cons of ceramics. This procedure involves reduction of size of the tooth to accommodate the crown. Briefly, the aim of the procedure is to achieve good retention of restoration on crown, resistance of tooth to fracture, facilitate good chewing, protection of dental pulp while maintaining conservation of tooth structure. Unlike metals that are strong in thin sections, ceramics are strong only in bulk. Hence, when a particular region of crown is involved in mastication, more of tooth material is reduced at that particular region, called functional cusp bevel, to provide bulk to the ceramic. In such cases, amount of tooth reduction will be the sum of clearance for metal and ceramic in PFM and only for ceramic in all ceramic crowns. This definitely projects the tooth conservation in all-ceramic crowns.

In order to overcome the disadvantages of PFMs, and to achieve closer tooth colour match, All-Ceramics were developed. Natural teeth always permit diffuse transmission and regular transmission. Prosthetic teeth must also possess a similar depth of translucency, which is realized in all-Ceramic restorations. Aluminous porcelain, Glass Ceramics, Castable, Machinable and Pressable Glass infiltrated, CAD-CAM and Cercon Zirconia system are few examples of all-ceramic systems.

Category 3: Crystalline-based systems with glass fillers:

Aluminous porcelain contains a glass matrix phase and at least 35 vol% of Alumina. It is a commonly used core ceramic and has a thin platinum foil when employed with all ceramic restorations. Aluminous core is stronger than feldspathic porcelain.

Construction of the Procera AllCeram core involves die production from the impression, digitising the geometry of coping using computer software, and transferring this information to a laboratory in Stockholm. The coping is produced by a process, which involves sintering 99.5% pure alumina at 1600–1700 °C such that it is completely densified. The coping is then returned to the dental laboratory for building in the crown's aesthetics using compatible feldspathic glasses. The flexural strength of the Al_2O_3 core materials is in the region of 700MPa, and thus similar to that achieved with the In-Ceram Zirconia.

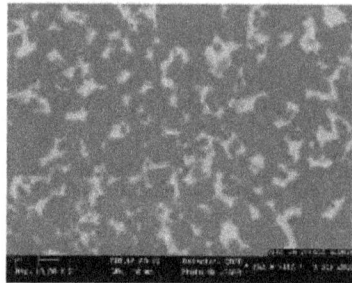
High density of In-Ceram powder.

Pure alumina cores are produced by Techceram Ltd. In this system the impression can be sent to Techceram Ltd, who will produce a special die onto which the alumina core is deposited using a thermal gun-spray technique. This process produces an alumina core with a density of 80–90%, which is subsequently sintered at 1170 °C to achieve optimum strength and translucency. The alumina coping is then returned to the dental laboratory, where the ceramist will develop the final contour and aesthetics using conventional feldspathic glasses. The clinical significance is the better translucency than glass alumina ceramics.

SEM image of In-Ceram surface.

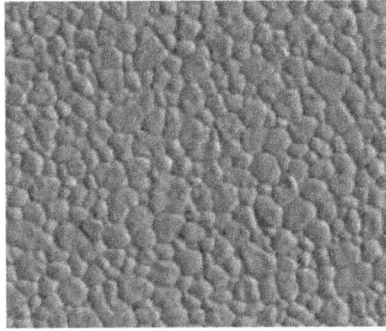

SEM image of Sintered Zirconia (Lava).

The use of modern Aluminous Crowns rose in the mid 1960's by McLean. The Nobel Biocare company (Sweden) introduced systems of pressing alumina onto a metal die, removing the pressed shape from the die and then sintering it. They are used to make cores to build up ceramic superstructures for dental implants (CeraOne®), and the second for conventional crowns (Procera®). Here, there is no glassy phase present between the particles. Feldspathic veneering porcelains are then fired onto this core to provide the necessary colour and form for the final restoration.

SEM Image of Procera Crown.

The Pressable glass ceramics involve pressure molding in the manufacture. Heated ceramic ingot is pressed through a heated tube into a mold, where the ceramic form cools and hardens to the shape of the mold, which is later recovered after cooling. Hot pressing usually occurs over a 45mins at a high temperature to produce the ceramic sub-structure. Then it is stained, glazed or coated by veneering porcelain, according to esthetic needs, which results in translucent ceramic core, moderately high flexural strength, excellent fit & excellent esthetics.

The 1980s witnessed the development of slip casting system using fine grained alumina. The cast alumina was initially sintered and then infiltrated with a Lanthana based glass. Onto this alumina core, a feldspathic ceramic could be baked to provide form, function and esthetics to the crown. Glass infiltrated ceramic is used as one of the 3 core ceramics namely, In-Ceram Spinell, In-Ceram Alumina and In-Ceram Zirconia. They use the technique of slipcast on a porous refractory die and heated in a furnace to produce a partially sintered coping or framework which is infiltrated with glass at 1100 °C for about 4 hrs to strengthen the slip-cast core. They possess high flexural strength and can be successfully cemented using any cement. CAD/CAM involves a technique wherein the internal surface is ground with diamond discs to the dimensions obtained from a scanned image of the preparation.

In-Ceram consists of a family of all-ceramic restorative materials. The family encompasses a range of strengths, translucencies, and fabrication methodology designed to cover the wide scope of all-ceramic restorations, including veneers, inlays, onlays, and anterior/posterior crowns and bridges. In-Ceram Spinell (alumina and magnesia matrix) is the most translucent with moderately high strength and used for anterior crowns. In-Ceram Alumina (alumina matrix) has high strength and moderate translucency and is used for anterior and posterior crowns. In-Ceram Zirconia (alumina and zirconia matrix) has very high strength and lower translucency and is used primarily for three-unit posterior bridges. In addition, these materials are supplied in a block form for producing milled restorations using a variety of machining systems.

In-Ceram is in a class called interpenetrating phase composites. They consist of at least two phases, which are intertwined and extend continuously from the internal to external surfaces. This class has better mechanical and physical properties relative to the individual components; a tortuous route through alternating layers of both components is required in order for these materials to break.

Interpenetrating phase materials are generally fabricated by first creating a porous matrix; in the case of In-Ceram, it would be a ceramic "sponge." The pores are then filled by a secondphase material, lanthanum-aluminosilicate glass, using capillary action to draw a liquid or molten glass into all the pores to produce the dense interpenetrating material.

The system met with great success as an alternative to conventional metal-ceramics. It uses a sintered crystalline matrix of a high-modulus material (85% of the volume) in which there is a junction of the particles in the crystalline phase. This is different from glasses or glassceramic materials in that these ceramics consist of a glass matrix with or without a crystalline filler in which there is no junction of particles (crystals). Slip casting may be used to fabricate the ceramic matrix, or it can be milled from a presintered block. Flexural strengths range from 350 MPa for spinell, 450 MPa for alumina, and up to 650 MPa for zirconia. Clinical studies support In-Ceram Alumina as an almost all purpose ceramic. InCeram Alumina had the same survival rates as porcelain-fused-to-metal restorations up to the first molar, with a slightly higher failure rate for the second molar. In-Ceram Zirconia should be used on molars only due to its very high opacity, which is not suitable for anterior esthetics. For anterior teeth, the alumina magnesia version of In-Ceram (called Spinell) is ideal due to its higher translucency.

Polycrystalline Solids

Solid-sintered monophase ceramics are formed by directly sintering crystals together without any intervening matrix to form a dense, air-free, glass-free, polycrystalline structure. Several processing techniques allow the fabrication of either solid-sintered aluminous oxide (alumina, Al_2O_3) or zirconium oxide (ZrO_2) framework. The first fully dense polycrystalline material for dental applications was Procera® AllCeram alumina, with a strength of approximately 600 MPa. The alumina powder is pressed and milled on a die and sintered at about 1600 °C, leading to a dense coping but with approximately 20% shrinkage.

The use of zirconia has increased rapidly in the past few years. This is Zirconia, partially stabilized with small amounts of other metal oxides. Partially stabilized zirconia allows production of reliable multiple-unit all-ceramic restorations for posterior teeth, since they produce high stress. Zirconia may exist in several crystal types (phases) depending on the addition of minor components. Typically for dental applications, about 3 wt% of yttria is added to pure zirconia.

Zirconia has unique physical characteristics that make it twice as strong and tough as alumina-based ceramics. Values for flexural strength range from approximately 900 MPa to 1100 MPa. There is no direct correlation between flexural strength (modulus of rupture) and clinical performance. Another important physical property is fracture toughness, which has been reported between 8 MPa and 10 MPa for zirconia. This is significantly higher than other dental ceramics. Fracture toughness is a measure of a material's ability to resist crack growth. Zirconia has the apparent physical properties to be used for multiple-unit anterior and posterior FPDs. Clinical reports on zirconia have not shown any problem with the framework, but have shown the chipping and cracking of porcelain. Using a slow-cooling protocol at the glaze bake to equalize the heat dissipation from zirconia and porcelain increased the fracture resistance of the porcelain by 20%. Zirconia may be in the form of blocks that are milled to create the frameworks (CAD/CAM). Mostly, they are fabricated from a porous block, milled oversized by about 25%, and sintered to full density in a 4 - 6 hours cycle. Alternatively, fully dense blocks are milled. However, this approach requires approximately 2 hours of milling time per unit whereas milling of the porous block necessitates only 30 to 45 minutes for a three-unit bridge.

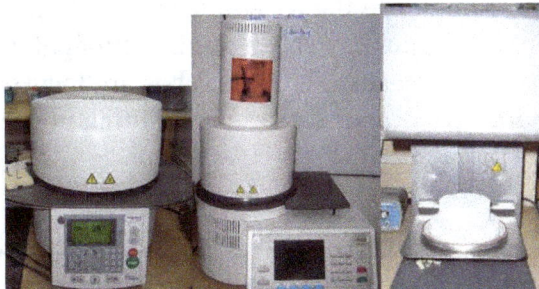
Dental Ceramic Furnaces.

Inlays and Onlays

Basically, Inlay is an indirect filling placed on teeth. Onlay is a more extensive filling (Involving more that one cusp in case of molars). They work in similar fashion.

These are made of many kinds of previously mentioned ceramics. They are fabricated by CAD/CAM technique.

Implants

Implant is basically a pillar/post placed into bone to act as a replacement to root of a tooth, on which a denture – fixed/removable is placed. Usually these are made from titanium and its alloys. Basic criteria for a material to be used as an implant is strength and bone biocompatibility. Recently Zirconia implants are introduced in to the market. With an increasing consciousness and fear for metal allergy, Zirconia implants are gaining momentum. Their advantage is their non allergic nature. In clinical studies, no difference in performance is reported. Their basic disadvantage over titanium is their brittleness. Also hydroxyapatite coated titanium implants are already into the market. These aim at better tissue response and osseointegration. Forces experienced by the implants are basically compressive and bending. Even though alumina has good mechanical properties, it is inert towards bone, hence not used in implant dentistry.

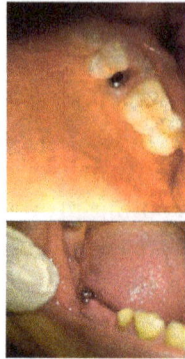

Picture of titanium implant in place - clinical picture.

Zirconia implants (Ceraroot etc) are compressed, sintered and milled to produce the necessary shape and surface texture. In case of dental implants, it is a proven fact that shape and surface texture influence the osseointegration. Right from introduction of implants in dentistry, lots of shapes have been tried. Lots of patented designs are found regarding these designs and surfaces. The researches are focused on chemistry, morphology and surface characteristics for betterment of implant success.

Braces

The science of braces – called Orthodontics, describes the corrective procedure as unaesthetic metamorphosis to an esthetic result. The aim of ceramic braces is to reverse the above description. The primary use of these brackets is esthetics. Ceramic braces as such do not experience masticatory forces, But are subjected to the orthodontic forces like sliding over wires, torquing etc. Due to its high wear coefficient, when wrongly placed can cause attrition of opposing teeth. They are basically polycrystalline alumina or Zirconia.

Ceramic braces, colour codes showing
the correct tooth to be cemented to.

Classification Based on Processing Techniques

For ease of understanding, classification according to processing techniques is advocated. Generally, there are:

1. Powder/liquid glass-based systems.

2. Pressable blocks of glass-based systems.

3. CAD/CAM systems.

This system is more relevant clinically. In spite of same chemistry and microstructure, the processing history determines the properties relevant to clinical performance. Specifically, machined blocks of materials have performed better than powder/liquid versions of the same material.

Powder/Liquid Systems

These can be conventional or slip casted. The conventional construction of a porcelain crown involves compaction, firing and glazing. Briefly compaction involves mixing of powder with water and binder to form a paste by spatulation, brush application, whipping or vibrating, that is aimed at compaction, which is painted over the die that is previously coated with platinum foil. This paste is made from different porcelain powders to mimic the esthetics of natural teeth. Usually, an opaque shade (to mask metal core), a dentin shade and then enamel shade is used. The enamel shade is selected from shade guide matched to patients actual tooth shade.

The objective of condensation techniques is to remove water, resulting in a more compact arrangement with high density of particles that reduces the firing shrinkage. The particle shape and size affect the handling characteristics of the powder and have influence on firing shrinkage. The binder helps to hold the fragile particles together in this so-called green state.

Firing initially involves slow heating of crown in the open entrance to the furnace, to drive off excess water before it forms steam that cracks the ceramic. Dried compact is placed in the furnace and the binders are burnt out. Some contraction occurs in this stage. When the porcelain begins to fuse, continuity is achieved at contact points between the powder particles. The material is still porous, and is usually referred to as being at the low bisque stage. As the higher temperature prevails for longer time, more fusion takes place as the molten glass flows between the particles, resulting in more compaction and filling the voids. A large contraction takes place during this phase (~ 20%), and the resultant material is apparently non-porous. The high shrinkage is caused by fusion of the particles during sintering, and resultant close contact between particles. Longer sintering will lead to pyroplastic flow and loss of form and will become highly glazed. A very slow cooling rate is employed to avoid cracking or crazing.

The furnaces can be programmed to automate these procedures. Vacuum-firing produces a denser porcelain than air firing, as air is withdrawn during the firing process, resulting in fewer voids and a stronger crown and more predictable shade. Areas of porosity in air fired porcelain alter the translucency of the crown, as they cause light to scatter. Also, air voids become exposed on grinding of the superficial layer, compromising esthetics by giving a rough surface finish.

Glazing is done to eliminate residual surface porosity that might encourage bacterial colonization and its sequel. Glazing results in surface that is smooth, shiny and impervious. To accomplish this, either low fusing glasses are applied to crown after construction and fused, or final firing is done under controlled condition that fuses the superficial layer to make it impervious.

With regard to slip casting, the "slip" is a homogenous dispersion of ceramic powder in water. The water pH adjustment creates a charge on the ceramic particles, which are coated with a polymer to cause the fine suspension in water. In the case of In-Ceram, the slip is applied on a gypsum die to form the underlying core for the ceramic tooth. The water is absorbed by porous gypsum, leading to packing of particles into a rigid network. The alumina core is then slightly sintered in a furnace to create an interconnected porous network. The lanthanum glass powder is placed on the core and glass becomes molten and flows into the pores by capillary action to produce the interpenetrating network. The last step in the fabrication involves application of aluminous porcelain on the core to produce the final form of the restoration. Other powder dispersions, such as those created with zirconia, may be poured into a gypsum mold that removes the water and leads to formation of homogeneous block of zirconia.

Pressable Ceramics

Pressed ceramic restorations are fabricated using a method described previously, similar to injection molding. Empress restorations and other materials with a similar leucite/glass structure are fabricated in this manner. Pressables may be used for inlays, onlays, veneers, and single-unit crowns.

CAD/CAM

Numerous ceramics have found their way into this system, due to its short processing time. Some of them are described here. Glass/Crystal ceramics are made from fine-grain powders, producing pore-free ceramics. This was the first material specifically produced for the CEREC system. It has an excellent history of clinical success for inlays, onlays, and anterior and posterior crowns. These blocks are available as monochromatic, polychromatic with stacked shades as in a layered cake, and in a form replicating the hand-fabricated crowns whereas an enamel porcelain is layered on top of dentin porcelain. Glass/Leucite is a feldspathic glass with approximately 45% leucite crystal component. Lithium disilicate is not initially fully crystallized, which improves milling time and decreases chipping risk from milling. The milled restoration is then heat-treated for 20 - 30 mins to crystallize the glass and produce the final shade and mechanical properties of the restoration. This crystallization changes the restoration from blue to a tooth shade. The microstructural and chemical composition is essentially the same. Framework Alumina are fabricated by pressing the alumina-based powder into a block shape. These blocks are only fired to about 75% density. After milling, these blocks are then infused with a glass in different shades to produce a 100% dense material, which is then veneered with porcelain. Porous Alumina frameworks may be fabricated from porous blocks of material. Pressing the alumina powder with a binder into molds produces the blocks. The blocks may be partially sintered to improve resistance to machining damage or used as pressed in a fully "green" state (unfired, with binder). The frameworks are milled from the blocks and then sintered to full density at approximately 1500 °C for 4 to 6 hours. The alumina has a fine particle size of about 1μm and strength of approximately 600 MPa and is designed for anterior and posterior single units, as well as anterior three-unit bridges. Porous Zirconia frameworks milled from porous blocks are fabricated similarly to alumina blocks. As is the case with the alumina block, the milled zirconia framework shrinks about 25% after a 4 - 6 hours cycle at approximately 1300 °C to 1500 °C. The particle size is about 0.1 μm to 0.5 μm.

CERAMIC ART

Ceramic art is art made from ceramic materials, including clay. It may take forms including artistic pottery, including tableware, tiles, figurines and other sculpture. Ceramic art is one of the arts, particularly the visual arts. Of these, it is one of the plastic arts. While some ceramics are considered fine art, as pottery or sculpture, some are considered to be decorative, industrial or applied art objects. Ceramics may also be considered artefacts in archaeology. Ceramic art can be made by one person or by a group of people. In a pottery or ceramic factory, a group of people design, manufacture and decorate the art ware. Products from a pottery are sometimes referred to as "art pottery". In a one-person pottery studio, ceramists or potters produce studio pottery.

Most traditional ceramic products were made from clay (or clay mixed with other materials), shaped and subjected to heat, and tableware and decorative ceramics are generally still made this way. In modern ceramic engineering usage, ceramics is the art and science of making objects from inorganic, non-metallic materials by the action of heat. It excludes glass and mosaic made from glass *tesserae*.

Venus of Dolni Vestonice, before 25,000 BCE.

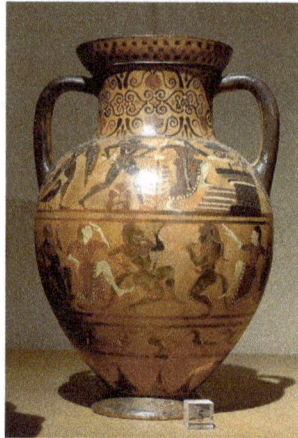

Etruscan: Diomedes and Polyxena, from the
Etruscan amphora of the Pontic group,

There is a long history of ceramic art in almost all developed cultures, and often ceramic objects are all the artistic evidence left from vanished cultures, like that of the Nok in Africa over 2,000 years ago. Cultures especially noted for ceramics include the Chinese, Cretan, Greek, Persian, Mayan, Japanese, and Korean cultures, as well as the modern Western cultures.

Elements of ceramic art, upon which different degrees of emphasis have been placed at different times, are the shape of the object, its decoration by painting, carving and other methods, and the glazing found on most ceramics.

Materials

Different types of clay, when used with different minerals and firing conditions, are used to produce earthenware, stoneware, porcelain and bone china (fine china).

- Earthenware is pottery that has not been fired to vitrification and is thus permeable to water. Many types of pottery have been made from it from the earliest times, and until the 18th century it was the most common type of pottery outside the far East. Earthenware is often made from clay, quartz and feldspar. Terracotta, a type of earthenware, is a clay-based unglazed or glazed ceramic, where the fired body is porous. Its uses include vessels (notably flower pots), water and waste water pipes, bricks, and surface embellishment in building construction. Terracotta has been a common medium for ceramic art.

- Stoneware is a vitreous or semi-vitreous ceramic made primarily from stoneware clay or non-refractory fire clay. Stoneware is fired at high temperatures. Vitrified or not, it is nonporous; it may or may not be glazed. One widely recognised definition is from the Combined Nomenclature of the European Communities, a European industry standard states "Stoneware, which, though dense, impermeable and hard enough to resist scratching by a steel point, differs from porcelain because it is more opaque, and normally only partially vitrified. It may be vitreous or semi-vitreous. It is usually coloured grey or brownish because of impurities in the clay used for its manufacture, and is normally glazed."

- Porcelain is a ceramic material made by heating materials, generally including kaolin, in a kiln to temperatures between 1,200 and 1,400 °C (2,200 and 2,600 °F). The toughness, strength and translucence of porcelain, relative to other types of pottery, arises mainly from vitrification and the formation of the mineral mullite within the body at these high temperatures. Properties associated with porcelain include low permeability and elasticity; considerable strength, hardness, toughness, whiteness, translucency and resonance; and a high resistance to chemical attack and thermal shock. Porcelain has been described as being "completely vitrified, hard, impermeable (even before glazing), white or artificially coloured, translucent (except when of considerable thickness), and resonant". However, the term *porcelain* lacks a universal definition and has "been applied in a very unsystematic fashion to substances of diverse kinds which have only certain surface-qualities in common".

- Bone china (fine china) is a type of soft-paste porcelain that is composed of bone ash, feldspathic material, and kaolin. It has been defined as *ware with a translucent body* containing a minimum of 30% of phosphate derived from animal bone and calculated calcium phosphate. Developed by English potter Josiah Spode, bone china is known for its high levels of whiteness and translucency, and very high mechanical strength and chip resistance. Its high strength allows it to be produced in thinner cross-sections than other types of porcelain. Like stoneware it is vitrified, but is translucent due to differing mineral properties. From its initial development and up to the later part of the twentieth century, bone china was almost exclusively an English product, with production being effectively localised in Stoke-on-Trent. Most major English firms made or still make it, including Mintons, Coalport, Spode, Royal Crown Derby, Royal Doulton, Wedgwood and Worcester. In the UK, references to "china" or "porcelain" can refer to bone china, and "English porcelain" has been used as a term for it, both in the UK and around the world. Fine china is not necessarily bone china, and is a term used to refer to ware which does not contain bone ash.

Surface Treatments

China Painting

China painting, or porcelain painting is the decoration of glazed porcelain objects such as plates, bowls, vases or statues. The body of the object may be hard-paste porcelain, developed in China in the 7th or 8th century, or soft-paste porcelain (often bone china), developed in 18th-century Europe. The broader term ceramic painting includes painted decoration on lead-glazed earthenware such as creamware or tin-glazed pottery such as maiolica or faience. Typically the body is first fired in a kiln to convert it into a hard porous biscuit. Underglaze decoration may then be applied, followed by ceramic glaze, which is fired so it bonds to the body. The glazed porcelain may then be decorated with overglaze painting and fired again at a lower temperature to bond the paint with the glaze. Decoration may be applied by brush or by stenciling, transfer printing, lithography and screen printing.

Slipware

Slipware is a type of pottery identified by its primary decorating process where slip is placed onto the leather-hard clay body surface before firing by dipping, painting or splashing. Slip is an aqueous suspension of a clay body, which is a mixture of clays and other minerals such as quartz, feldspar and mica. A coating of white or coloured slip, known as an engobe, can be applied to the article to improve its appearance, to give a smoother surface to a rough body, mask an inferior colour or for decorative effect. Slips or engobes can also be applied by painting techniques, in isolation or in several layers and colours. Sgraffito involves scratching through a layer of coloured slip to reveal a different colour or the base body underneath. Several layers of slip and/or sgraffito can be done while the pot is still in an unfired state. One colour of slip can be fired, before a second is applied, and prior to the scratching or incising decoration. This is particularly useful if the base body is not of the desired colour or texture.

Terra Sigillata

In sharp contrast to the archaeological usage, in which the term *terra sigillata* refers to a whole class of pottery, in contemporary ceramic art, 'terra sigillata' describes only a watery refined slip used to facilitate the burnishing of raw clay surfaces and used to promote carbon smoke effects, in both primitive low temperature firing techniques and unglazed alternative western-style Raku firing techniques. Terra sigillata is also used as a brushable decorative colourant medium in higher temperature glazed ceramic techniques.

Forms

Studio Pottery

Studio pottery is pottery made by amateur or professional artists or artisans working alone or in small groups, making unique items or short runs. Typically, all stages of manufacture are carried out by the artists themselves. Studio pottery includes functional wares such as tableware, cookware and non-functional wares such as sculpture. Studio potters can be referred to as ceramic artists, ceramists, ceramicists or as an artist who uses clay as a medium. Much studio pottery is tableware or cookware but an increasing number of studio potters produce non-functional or sculptural items. Some studio potters now prefer to call themselves ceramic artists, ceramists or simply artists. Studio pottery is represented by potters all over the world.

Tile

A tile is a manufactured piece of hard-wearing material such as ceramic, stone, metal, or even glass, generally used for covering roofs, floors, walls, showers, or other objects such as tabletops. Alternatively, tile can sometimes refer to similar units made from lightweight materials such as perlite, wood, and mineral wool, typically used for wall and ceiling applications. In another sense, a "tile" is a construction tile or similar object, such as rectangular counters used in playing games.

Upper part of the mihrab decorated with lusterware tiles (dating from the 9th century) in the Mosque of Uqba also known as the Great Mosque of Kairouan, Tunisia

Tile, Hopi Pueblo (Native American).

Tiles are often used to form wall murals and floor coverings, and can range from simple square tiles to complex mosaics. Tiles are most often made of ceramic, typically glazed for internal uses and unglazed for roofing, but other materials are also commonly used, such as glass, cork, concrete and other composite materials, and stone. Tiling stone is typically marble, onyx, granite or slate. Thinner tiles can be used on walls than on floors, which require more durable surfaces that will resist impacts.

Figurines

Group with lovers, modelled by Franz Anton
Bustelli, Nymphenburg porcelain, 1756.

A figurine (a diminutive form of the word *figure*) is a statuette that represents a human, deity, mythical creature, or animal. Figurines may be realistic or iconic, depending on the skill and intention of the creator. The earliest were made of stone or clay. In ancient Greece, many figurines were made from terracotta. Modern versions are made of ceramic, metal, glass, wood and plastic.

Figurines and miniatures are sometimes used in board games, such as chess, and tabletop role playing games. Old figurines have been used to discount some historical theories, such as the origins of chess.

Tableware

Tableware is the dishes or dishware used for setting a table, serving food and dining. It includes cutlery, glassware, serving dishes and other useful items for practical as well as decorative purposes. Dishes, bowls and cups may be made of ceramic, while cutlery is typically made from metal, and glassware is often made from glass or other non-ceramic materials. The quality, nature, variety and number of objects varies according to culture, religion, number of diners, cuisine and occasion. For example, Middle Eastern, Indian or Polynesian food culture and cuisine sometimes limits tableware to serving dishes, using bread or leaves as individual plates. Special occasions are usually reflected in higher quality tableware.

Terracotta (Artworks)

In addition to being a material, "terracotta" also refers to items made out of this material. In archaeology and art history, "terracotta" is often used to describe objects such as statures, and figurines not made on a potter's wheel. A prime example is the Terracotta Army, a collection of man-sized terracotta sculptures depicting the armies of Qin Shi Huang, the first Emperor of China. It is a form of funerary art buried with the emperor in 210–209 BCE and whose purpose was to protect the emperor in his afterlife.

French sculptor Albert-Ernest Carrier-Belleuse made many terracotta pieces, but possibly the most famous is The Abduction of Hippodameia depicting the Greek mythological scene of a centaur kidnapping Hippodameia on her wedding day. American architect Louis Sullivan is well known for his elaborate glazed terracotta ornamentation, designs that would have been impossible to execute in any other medium. Terracotta and tile were used extensively in the town buildings of Victorian Birmingham, England.

Ceramics Museums and Museum Collections

100 BCE – 250 CE.   Ancient Egyptian.

A ceramics museum is a museum wholly or largely devoted to ceramics, normally ceramic artworks, whose collections may include glass and enamel as well, but will usually concentrate on pottery, including porcelain. Most national ceramics collections are in a more general museum covering all the arts, or just the decorative arts, but there are a number of specialized ceramics museums, some concentrating on the production of just one country, region or manufacturer. Others have international collections, which may concentrate on ceramics from Europe or East Asia, or have global coverage.

Ceramic goblet from Navdatoli,
Malwa, India, 1300 BCE; Malwa culture.

In Asian and Islamic countries ceramics are usually a strong feature of general and national museums. Also most specialist archaeological museums, in all countries, have large ceramics collections, as pottery is the commonest type of archaeological artifact. Most of these are broken shards however.

A funerary urn in the shape of a "bat god" or a jaguar, from Oaxaca,
Mexico, dated to 300–650 CE. Height: 9.5 in (23 cm).

Outstanding major ceramics collections in general museums include The Palace Museum, Beijing, with 340,000 pieces, and the National Palace Museum in Taipei city, Taiwan (25,000 pieces); both are mostly derived from the Chinese Imperial collection, and are almost entirely of pieces from China. In London, the Victoria and Albert Museum (over 75,000 pieces, mostly after 1400 CE) and British Museum (mostly before 1400 CE) have very strong international collections. The Metropolitan Museum of Art in New York and Freer Gallery of Art in Washington DC (12,000, all East Asian) have perhaps the best of the many fine collections in the large city museums of the United States. The Corning Museum of Glass, in Corning, New York, has more than 45,000 glass objects.

Luca della Robbia, Virgin and
Child with John the Baptist.

18th century cocklestove in the
Catherine Palace, St Petersburg.

APPLICATIONS OF TECHNICAL CERAMICS IN TRANSPORTATION

Over the last several years, a surge in the production of novel ceramic materials, as well as advancements in the processing techniques for these materials, has allowed researchers to control and manipulate the specific microstructures of technical ceramics for a wide variety of purposes. In doing so, the application of these newly developed ceramic materials into the energy, environment and transportation industries has subsequently emerged as a promising technology that is expected to address global environmental and transportation concerns. As various industries, particularly the transportation sect, continue to utilize the unique properties of novel ceramic materials, some of which include low thermal expansion, robustness and high-temperature stability, this economic growth is expected to continue to rise significantly over the next several years.

Technical Ceramics

Ceramics are typically made up of commonly used and readily available materials such as carbon, silicon, oxygen, and nitrogen that, when consolidated under high-temperatures and pressures, can form ceramic materials that are used for a wide variety of products ranging from household to industrial applications. In comparison, technical ceramics, which are also referred to as engineered or high-performance ceramics, typically originate from more sophisticated compounds including aluminas, carbides, nitrides, borides, and zirconia2. Technical ceramics have traditionally been utilized for various electronic components including capacitors, resistors, semiconductor tools, and engine parts.

Fuel and Economic Effects of Ceramics in Transportation

Since the early 1920's, ceramics have been incorporated into various automobile components including spark plug insulators and glass windows. As research in this area continued to develop into the early 1980's, researchers found that technical ceramics were promising materials for the development of advanced engines for motor vehicles such as adiabatic diesel engines, gas turbines, and Stirling engines. Some of the most recent advancements in this area have found that since the high-temperature environments of automotive engines require highly durable materials to withstand these conditions, technical ceramics are able to provide automobiles with the necessary components to ensure peak engine performance while also extending the overall lifespan of all engine components.

Some of the most recent advancements in this area have involved incorporating ceramic coatings to diesel engine combustion chambers in an effort to reduce the heat that passes from in-cylinder to the engine cooling system. In fact, the dramatic ability of these ceramic coatings to reduce heat conductance within these internal combustion engines is expected to completely eliminate the need for engine cooling systems within these vehicles. Ceramic coated diesel engines have also been shown to reduce ignition delay during the initiation of applied engines as a result of the low heat rejection of these materials, thereby allowing a virtually silent engine operation. Additionally, ceramic coated internal combustion engines have also been shown to significantly reduce the amount of soot and carbon monoxide emissions from applied engines.

Composite Ceramics in Aviation

The turbofan engines that are currently used in most airplanes typically generate thrust be expelling rapidly moving hot gases from their core. In an effort to increase the efficiency of turbofan engines, Connecticut-based engine-maker Pratt & Whitney of United Technologies have developed the PurePower engines that are specifically designed for single-aisle jets. Although numerous engine parts within jets, including turbine blades, are currently coated with ceramic materials that allow engines to withstand temperatures as high as 1,500 °C[4], these coatings are often susceptible to spalling off and even reducing the efficiency of applied blades. What is particularly unique about the PurePower engines is that ceramic fibers have been used to reinforce the ceramic material that comprises this engine. Researchers expect that as commercial jets follow the trend in incorporating ceramic composite engines, fuel efficiency will subsequently increase as the overall engine weight will decrease by as much as 30%.

CERAMIC ARMOUR

Ceramic armor is armor used by armored vehicles and in personal armor for its attenuative properties. Ceramics provide projectile resistance through their high hardness and compressive strength and are often used in applications where weight is a limiting factor due to their lightweight nature relative to metals commonly used in armors. Most commonly alumina, boron carbide, silicon carbide, and titanium diboride ceramics are used in armor but other ceramics are used.

Ceramic armor has been in use by the US military since the Vietnam war although the first testing that demonstrated the potential ceramics had was in 1918. Major Neville Monroe-Hopkins found that by adding a thin layer of enamel, the ballistic properties of steel were greatly increased. Helicopters frequently came under small arms fire from small scale assaults which put the crews in greater danger. In 1965, ceramic body armor was given to the crews as well as 'hard-faced composite' armor kits placed within the pilot seats to offer better protection . Building off of this, the following year, monolithic ceramic vests and airframe-mounted armor panels were employed. These improvements are estimated to have decreased fatalities by 53% and the incident of non-fatal injuries by 27% in "Huey" helicopters. This was the first battle use of ceramic armor by any military.

The effectiveness of ceramic armour was further demonstrated in Desert Storm, where not a single British Army Challenger tank was lost to enemy tank fire. However, one was destroyed by friendly fire on March 25, 2003 killing two crew members after a HESH round detonated on the commander's hatch causing high velocity fragments to enter the turret. Chobham-type armour is currently in its third generation and is used on modern western tanks such as the British Challenger 2 and the American M1 Abrams.

Design

Ceramic armor comes in a variety of designs ranging from monolithic plating to systems employing three dimensional matrices. One of the first patents of ceramic armor was filed in 1967 by the Goodyear Aerospace Corp. It consisted of alumina ceramic spheres embedded into a thin aluminum sheets. These sheets were layered atop each other such that the spheres of other layers would fall within the spaces between spheres of the surrounding layers in a manner similar to a body-centered cubic packing structure. The remaining gaps were then filled with a polyurethane foam and the entire system was then given a thick aluminum backing to hold it together. This development demonstrated the effectiveness of matrix based design and thus spurred the development of other matrix based systems. Other matrix based designs can be included however, the main theme among the designs is the use of a ceramic based system with a backing composed of some non-armor dedicated alloy. Many of these designs can include systems employing cylindrical, hexagonal, or spherical ceramic pieces. Monolithic plate armor is also available. These come in the form of single plates of an advanced ceramic slipped into a traditional ballistic vest in place of a steel plate.

Mechanism

Hard ceramic materials defeat the kinetic energy projectile by shattering it into pieces, decreasing the penetration ability of projectile. In case of HEAT rounds the shattered ceramic fragments destroy the geometry of the metal jet generated by the shaped charge, greatly diminishing the penetration. Ceramic materials cannot be used as a stand alone for armour applications because of its shattering effect due to their inherent nature of brittleness. Therefore, in order to prevent the shattered ceramic and projectile pieces from further damaging the protected system, ceramic materials should always be supported by a ductile backing with metallic or polymeric composite materials. Another advantage of using backing material is to improve the ballistic performance of ceramics by preventing its premature failure.

Applications

Personnel

Ceramic plates are commonly used as inserts in soft ballistic vests. Most ceramic plates used in body armor provide National Institute of Justice Type III protection, allowing them to stop rifle bullets. Ceramic plates are a form of composite armor. Insert plates may also be manufactured from steel or ultra high molecular weight polyethylene.

A ceramic plate is usually slipped into the outer layer of a soft armor vest. There may be two plates, one in the front and one in the back, or one universal plate on either front or back. Some vests permit the usage of small plates on the sides for additional protection.

Ceramic plates issued by the United States military are called Enhanced Small Arms Protective Inserts (ESAPI).

The approximate weight for one NIJ Type III plate is 4 to 8 pounds (1.8–3.6 kg) for the typical size of 10" by 12". There are other types of plates that come in different sizes and offer different levels of protection. For example, the MC-Plate (maximum coverage plate) offers 19% more coverage than a standard ceramic plate.

Ceramic body armour plates.

Ceramic materials, materials processing and progress in ceramic penetration mechanics are significant areas of academic and industrial activity. This combined field of ceramics armor research is broad and is perhaps summarized best by The American Ceramics Society. ACerS has run an annual armor conference for a number of years and compiled a proceedings 2004–2007. An area of special activity pertaining to vests is the emerging use of small ceramic components. Large torso sized ceramic plates are complex to manufacture and are subject to cracking in use. Monolithic plates also have limited multi hit capacity as a result of their large impact fracture zone These are the motivations for new types of armor plate. These new designs use two and three dimensional arrays of ceramic elements that can be rigid, flexible or semi-flexible. Dragon Skin body armor is one these systems, although it has failed numerous tests performed by the US Army, and has been rejected. European developments in spherical and hexagonal arrays have resulted in products that have some flex and multi-hit performance. The manufacture of array type systems with flex, consistent ballistic performance at edges of ceramic elements is an active area of research. In addition advanced ceramic processing techniques arrays require adhesive assembly methods. One novel approach is use of hook and loop fasteners to assemble the ceramic arrays.

HIGH PRESSURE FUEL SYSTEMS

Diesel fuel systems are facing increased demands as engines with reduced emissions are developed. Injection pressures have increased to provide finer atomization of fuel for more efficient combustion, figure. This increases the mechanical loads on the system and requires tighter clearances between plungers and bores to prevent leakage. At the same time, fuel lubricity has decreased as a byproduct of reducing the sulfur levels in fuel. Contamination of fuel by water and debris is an ever-present problem. For oil-lubricated fuel system components, increased soot loading in the oil results in increased wear rates. Additionally, engine manufacturers are lengthening warranty periods for engines and systems. This combination of factors requires the development of new materials to counteract the harsher tribological environment.

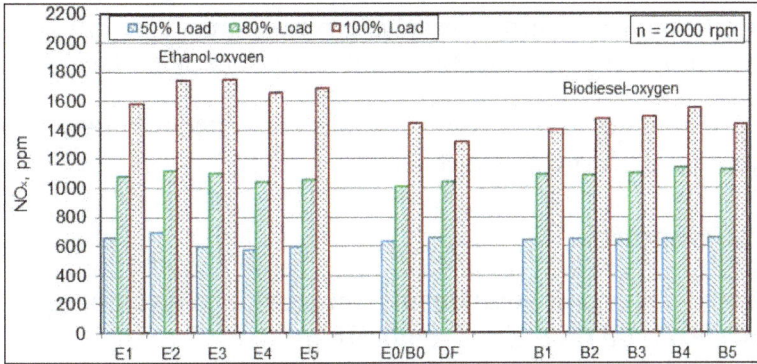

In highly loaded fluid-lubricated interfaces, boundary lubrication conditions often exist. In this situation the two surfaces are not completely separated by the fluid film and asperity contact occurs. Very high contact pressures exist at these points of contact. In the case of metal-on-metal contact, this can result in the micro-welding of asperities due to the similar metallic atomic bonding structure. Micro-welding results in material transfer from one surface to the other and a roughening of both surfaces, which further increases the asperity contact and frictional heat generation, eventually resulting in gross material transfer (adhesive wear), sticking, and seizing of the mating components.

Due to dissimilar atomic bonding, ceramic materials significantly reduce micro-welding of asperities and the cycle of degradation that takes place with steel-on-steel contact. Since ceramic materials are much harder than many tool steels, abrasive wear is also reduced. These characteristics increase the contact pressure/sliding velocity (PV) capability of a ceramic-steel interface.

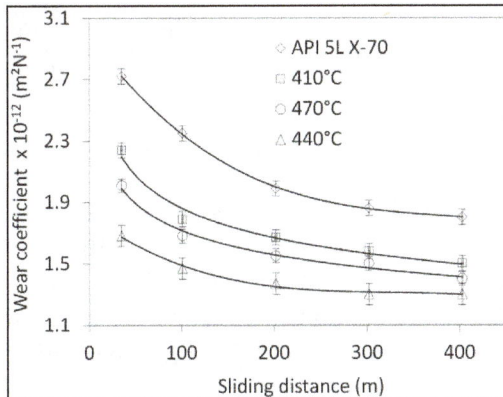

Evidence of this benefit was observed by Cummins in the mid-1980's with FALEX 1 rig testing which compares steel-steel and silicon nitride-steel combinations in sliding contact. Dramatic wear reduction was observed even in used lubricating oil containing high levels of soot.

Ceramics surfaces can be obtained as solid ceramic components or as coatings, such as titanium nitride (TiN) or amorphous carbon, on a steel substrate. Coating systems are the best choice for certain applications, but coating processes are often difficult to control and can be expensive. Coatings are also typically only a few microns thick. Once a coating is worn through to the substrate, adhesive wear again becomes a problem.

Solid ceramic components have become a viable, cost-effective alternative to coatings and provide superior performance to steel.

Applications/Benefits

The benefits that solid ceramic components can provide to fuel systems are illustrated through several successful production applications.

Fuel Injector Link

The first known application of a ceramic component in a heavy-duty diesel fuel system was the silicon nitride link used in the Cummins STCÔ unit injector, Figure. This component was introduced to improve the tolerance of the injector to higher levels of soot in the lubricating oil, as a result of changes required to meet the 1988 U.S. emission reductions. From previous wear testing, it was known that silicon nitride would provide improved wear resistance. Engine testing proved the performance increase.

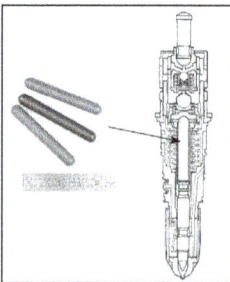

Silicon nitride injector links
and Cummins STCÔ unit injector.

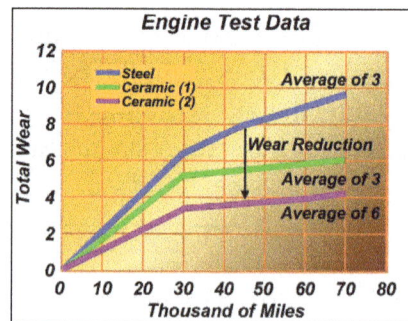

Engine test data
on silicon nitride link.

By replacing this one part, the total wear rate of the injector train was reduced significantly, more than doubling the mileage until adjustment was required. Additional changes to the injector increased this adjustment interval further. In the 11 years since introduction, 2 million links have been put into service without a single reported incident. The links are often re-used when the injectors are rebuilt, further evidence of the wear resistance of the ceramic link.

Fuel Pump Roller

Stanadyne's DS 50 electronically controlled fuel pump is used on the General Motors 6.5L diesel engine, which is installed in full-size pickups and vans. Stanadyne began seeing higher rates of pump distress after the federally mandated reduction in diesel fuel sulfur levels in October 1993. The steel rollers, which are fuel lubricated, were scuffing and seizing at an increased rate. The rollers experience both sliding and rolling contact in this application. Several years previously, Stanadyne had tested silicon nitride rollers for operation with kerosene and were familiar with this potential solution.

Stanadyne had worked on improving the form and surface finish of the steel rollers in an attempt to increase the lubricating film thickness and thereby reduce the boundary lubrication conditions. With the silicon nitride roller, form and surface finish tolerances could be relaxed, as the silicon nitride-on-steel combination was more tolerant to asperity contact.

Stanadyne DS 50 fuel pump.

Silicon nitride rollers.

The improved performance of the silicon nitride rollers was proven through rigorous abuse and durability testing. Abuse testing included high fuel temperature and pressure, overspeed condition, water in fuel, debris in fuel, and testing with low lubricity fuels. The improved performance has been demonstrated conclusively in the field. In the four years that the silicon nitride roller has been used in production, there has not been a single reported pump failure due to roller sticking or seizing.

High Pressure Check Balls

Check balls are another vulnerable fuel system component that seal pulsating, high fuel pressures. The combination of high frequency, high-load impact of the ball against the seat combined with the poor lubrication environment leads to deformation and wear of the ball and seat. The resulting fuel leakage can result in poor idle control, inability to turn the engine off, and other fuel system performance problems.

Silicon nitride has been proven to be a superior material for this application. With less than half the weight of steel, impact forces are reduced. The high hardness of the balls minimizes the deformation and causes the seat to conform to the ball for an improved seal. Reduced adhesive wear prevents transfer of material. Resistance to corrosion and being nonmagnetic are additional useful properties. Because the silicon nitride balls are widely used in the bearing industry, they are readily available in a variety of sizes at reasonable cost.

Incorporation of silicon nitride check balls in Cummins CELECTÔ injectors in 1992 improved the injector performance. Other major fuel system manufacturers are also using silicon nitride check balls in both diesel and gasoline systems.

High Pressure Pumping Plungers

The use of advanced ceramic materials for high-pressure pumping plungers has dramatically increased the reliability of Cummins' fuel systems in recent years. Cummins had experienced a certain level of plunger scuffing and seizing since the introduction of the CELECTÔ fuel system in 1990. The rate of incidents was observed to increase in winter months in northern climates when the use of DF1 was increased. Water in the fuel was also found to initiate scuffing and seizing. The problem became more widespread with sulfur reduction in diesel fuel in late 1993. Various attempts to improve the situation, such as improvements in bore and plunger geometry, surface treatments, different steel materials, etc. met with limited success.

Due to the low thermal expansion coefficient of silicon nitride, this material could not be used in this application. A tight clearance between the plunger and bore had to be maintained over the operating temperature range to prevent fuel leakage. Zirconia was determined to be a more suitable material. Cummins was familiar with zirconia technology as a result of work that was sponsored by the DoD in the early 1980s. Cummins was the project manager of a program to work with U.S. suppliers to develop advanced zirconia materials capability in this country. Figure shows a sectioned schematic of the Cummins CELECTÔ injector with the pumping plungers visible.

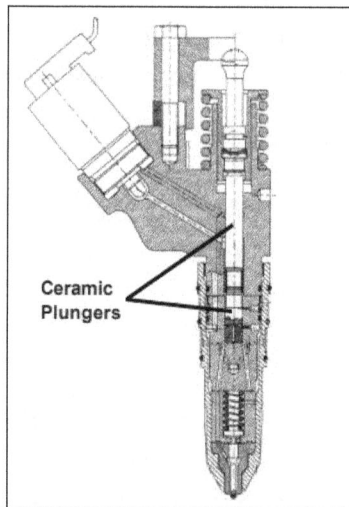

Cummins CELECTO injector.

Testing of injectors with ceramic plungers provided dramatic evidence of improvement. With 1% water in fuel, steel plungers would seize in 10 hours or less. Zirconia plungers operated more than 300 hours with 2% water in fuel without failure. Plungers were also tested with an angled top surface to purposely increase the side loading forces. There was no affect on performance. Other tests included overspeed and overpressure conditions and alternative fuel (low lubricity) testing. In all cases, the zirconia plungers far exceeded the performance of the steel plungers. A comparison test to determine injector bore wear when using plungers of different materials found zirconia to be far superior to uncoated steel and titanium nitride coated steel.

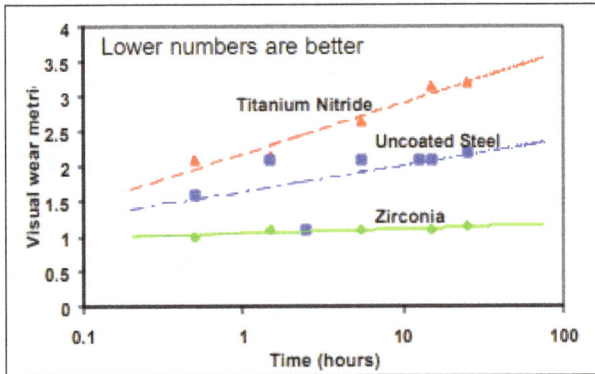

Injector bore wear as a function
of plunger material and time.

The improvement provided by the zirconia plungers was immediate and dramatic in production. Occurrences of plunger sticking were essentially eliminated with the introduction of the first zirconia plunger, Figure. A dramatic improvement in injector reliability resulted as the use of zirconia plungerequipped injectors became widespread.

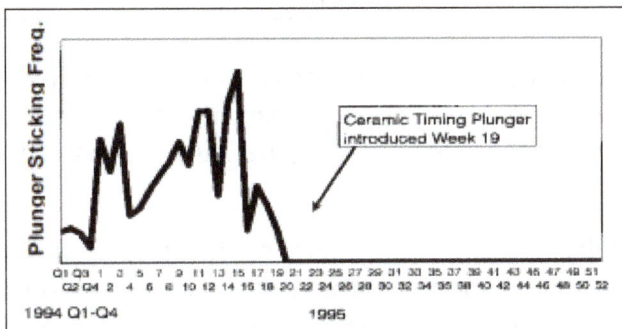

Plunger sticking frequency.

The excellent experience with the CELECTO™ injector led Cummins to incorporate zirconia plungers into the new CAPSO™ common rail fuel system early in the development program, Figure. This marked a milestone, as it was the first time that a diesel engine company had introduced a ceramic component before a system was in commercial production. All prior ceramic applications had been retrofits of an existing steel component to resolve some wear or scuffing issue in a production system.

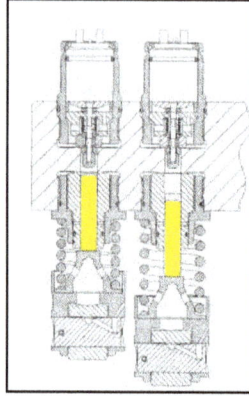

CAPSO high pressure pump
section showing ceramic plungers.

Ceramic materials have proven to be a valuable technology for improving the reliability and durability of high-pressure diesel fuel systems in the more challenging environment brought on by the drive to produce engines with lower emission levels. The reduction in adhesive wear between ceramic and steel components in boundary lubrication conditions has resulted in dramatic reductions in seizing, sticking and wear of diesel fuel system components that operate with reduced lubricity and/or contaminated fuels and oils. The increased capability of these materials allow fuel system developers to design more capable systems that can operate as-designed for longer periods of time. Continuing efforts to develop improved materials and machining processes and the confidence gained with these successful applications will allow ceramics to be applied in new ways and at lower cost.

AUTOMOTIVE ENGINEERING

Motorized vehicles are manufactured in series production with a high degree of automation. With the variety of technical equipment, the numerous design variants and the high utility value for the customers, they are hardly surpassed by other technical products.

A large number of industries and technologies are involved in the manufacture of motorized vehicles: machine tool manufacturing, glass industry, plastics industry, ceramic and chemical industry, electrical engineering and electronics industries, textile industry, surface finishing and environmental engineering, just to name a few examples.

In a motorized vehicle, the following main component assemblies can be defined:

- Engine,

- Power transmission/drive train,

- Chassis,

- Car body,

- Vehicle electrics/electronics.

Every component assembly has specific technical requirements for the materials used. The selection of certain materials is guided by the goal to maximize energy and cost efficiency combined with acceptable reliability.

As in the majority of applications in engineering, in motorized vehicles too, components made of high-quality technical ceramics are used to reliably meet requirements that materials on metal or plastic basis are hardly able to fulfil.

Often, the mainly dense-sintered ceramic materials make economic realization of requirements possible in first place. The spark plug with its electrical insulator made of Al_2O_3 ceramic is a historical example or the l-sensor with doped ZrO_2 as an electric conductor provides an example from more recent years.

Application in motorized vehicles demands from components made of technical ceramics high reliability and cost efficiency in long-term operation. The application-specific requirements are therefore focussed on the following properties:

- Mechanical strength,

- Density,

- Achievable geometric precision and edge stability,

- Tribological properties, e.g. coefficient of friction and abrasive behaviour even in emergency conditions,

- Dimensional stability with changing thermal and mechanical loads,

- Resistance to high temperatures and sudden temperature changes,

- Insulating capacity and thermal conductivity,

- Chemical corrosion resistance,

- Electrical insulation and electrical conductivity,

- Dielectric properties,

- Magnetic properties,

- Suitability for thin and thick film technologies,

- Possibility to produce force-fit, form-fit and adhesively bonded ceramic-ceramic and ceramic-metal joints.

Today, the manufacturers of motor vehicles use monolithic ceramic materials, composites, piezoceramics and magnetoceramics on oxide and non-oxide basis. The components made of these materials are often optimized for the specific application. As a result, they achieve high reliability and long-term durability in everyday operation.

In motor vehicles too, the typical characteristic of the applications of ceramic materials is their existence in positions within component assemblies that are generally not visually accessible. One exception in this connection is the brake disk made of a non-oxide ceramic fibre composite, the use of which brings key technical benefits compared with conventional brake disks, like, for example, high wear resistance and consequently an unusually long service lifetime in operating conditions.

APPLICATION OF CERAMICS IN THERMAL INSULATION APPLICATION

For most of people, ceramics are synonymous with hard, density ceramic tableware or wall tile, but modern technologies enable people to produce highly porous foam ceramics, intended for use as energy-efficient thermal insulation in blast furnaces. The new developed porous ceramic material possesses a unique combination of thermophysical and thermomechanical properties, as a result of high porosity, demonstrating outstanding operating characteristics with working temperatures of up to 1800 °C.

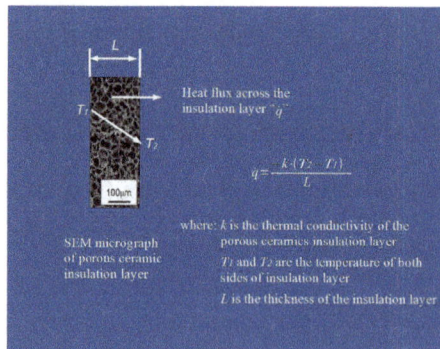

Heat flux across the insulation layer "q"

$$q = \frac{k \cdot (T_2 - T_1)}{L}$$

where: k is the thermal conductivity of the porous ceramics insulation layer

T_1 and T_2 are the temperature of both sides of insulation layer

L is the thickness of the insulation layer

SEM micrograph of porous ceramic insulation layer

Heat transfer in porous ceramics is a complicated process which comprises conduction, convection and radiation, however in many cases, the latter two items are usually much smaller than conduction so they can be neglected.

Heat flux across the porous ceramic layer "q" (W/m^2) is related to the temperature difference between two sides of the layer "$T_2 - T_1$" (K), the thickness of the layer "L" (m) and the thermal conductivity of the porous ceramic layer "k" ($W/(mK)$). The thermal conductivity "k" is a parameter to estimate the heat transfer capability of a material: if k is larger, we say the material is a good thermal conductor; when k is small on the other hand, the material is then regarded as thermal insulator.

Copper is a typical thermal conductor: k_{copper} = 401 $(W/(mK))$ at 25 °C. Air is usually treated as thermal insulated: k_{air} = 0.024 $(W/(mK))$ at 25 °C. The thermal conductivities of several common density ceramic material are list in the following Table.

Density Ceramic Material	Thermal Conductivity (W/(mK))
Cordierite	1 to 3
Zirconium oxide (ZrO_2)	2.5 to 3
Sintered silicon carbide (S-SiC)	100 to 140
Clay-bound silicon carbide (CB-SiC)	10
Aluminum oxide (Al_2O_3)	25

Research shows that the thermal conductivity of porous ceramics can be approximately expressed as a function of the porosity of the material:

$$k = k_0 \cdot e^{\left(\frac{-1.5 \cdot \varphi}{1-\varphi}\right)}$$

k_o is the thermal conductivity of density ceramics whose porosity is zero; φ is the porosity of porous ceramic material. According to the formula, the thermal conductivities of several ceramics at various porosity can be plotted as figure.

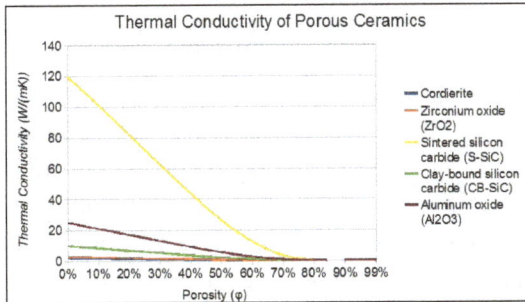

The figure shows even the thermal conductivities of density ceramics have larger difference, they become almost similar when their porosities are bigger than 75%. This provides boarder choices of material when designing thermal insulation. Benefits of porous ceramics for use as thermal insulation include:

1. longer life,

2. more uniform in firing atmosphere,

3. less shrinkage in insulation layer,

4. chemical inertness,

5. cost effective,

6. ceramic fibre free.

APPLICATION OF CERAMICS IN CHEMICALS AND PHARMACEUTICALS

Machines and equipment in chemical and pharmaceutical process engineering are often exposed to high stresses caused by temperature, pressure, corrosion and abrasion. Owing to the constantly rising requirements for the efficiency of the processes, the most frequently used metallic materials in the area in contact with the products are often exposed to such high stresses that they only reach uneconomically short service lifetimes in operation. The result is cost-driving downtime of machines and equipment.

In such cases, ceramic materials on oxide and non-oxide basis are often a suitable alternative for the designers. In this connection, an exact knowledge of the application conditions is crucial for selection of the right material.

The thermal strength of dense-sintered, high-purity oxide ceramics generally reaches the region of 2000 °C. Non-oxide ceramic materials are less thermally resistant on account of the restriction of their oxidation stability in an oxygen-containing atmospheres. Depending on the type of material, oxidation reactions only become noticeable above 1500 °C, such reactions can be slowed down substantially or even prevented completely with the formation of passivation layer. In non-oxidizing environment, these materials withstand a type-dependent temperature level, which reaches in the region of 2000 °C.

An economically significant issue in chemical process engineering is corrosion resistance of the materials in apparatus and equipment engineering when these are in contact with different highly concentrated acid and basic aqueous solutions of varying purity. In this respect, oxide and non-oxide ceramic materials have proven effective components over decades, which, apart from few exceptions, have achieved extraordinarily long service lifetimes in use in contact with such media even under tough conditions in numerous applications.

The behaviour of the ceramic materials exposed to corrosive load is determined mainly by their chemical composition and their microstructure. They can reach high corrosion resistance especially when they are formed by means of solid phase sintering and therefore have only a low content of grain boundary phase.

The following explanation shows, based on the example of Al_2O_3 ceramic, a simplified model explaining the background to corrosion processes and some possibilities to limit these.

On closer examination, various types of corrosion can be identified. Providing the crystalline Al_2O_3 base material is resistant, the focus is on the phase found on the grain boundaries of the microstructure. This intercrystalline material has not dissolved at all or only to a limited extent in the Al_2O_3 crystal during sintering of the ceramic, or it has been deposited during cooling from the sintering temperature. The chemical composition of this grain boundary phase and its structural state (amorphous or crystalline) then determine the corrosion resistance of the ceramic, as, of course, it has a completely different composition than the base crystal. Its chemical properties may promote corrosion processes, but also slow them down, for example, as a result of the formation of corrosion-resistant reaction products.

In the former case, a corrosion front will develop over time, which as a result of the dissolution of the grain boundary phase and the fallout of individual crystallites from the crystallite bond can lead to its complete destruction.

Such intercrystalline corrosion processes can be influenced by at least two principal measures and can even be halted completely:

- Adaption of the chemical composition of the grain boundary phase to the corrosive conditions.

- Reduction of the crystallite sizes of microstructure and as a result interruption or at least lengthening of the corrosion paths.

With the use of materials with 99,9 % purity and more, such measures can be usually avoided. However, this is a route associated with high costs as already the prices of such raw materials are according to current levels at least ten times higher than the Al_2O_3 raw materials with at least 99,8 % purity produced in annual volumes of 100 000 t and more.

With the technical means available today, it is no longer an insurmountable hurdle to produce a tailored Al_2O_3 ceramic optimized to combat specific corrosive loads. Providing all components of the corrosion system with the data on concentrations, pressure and temperature are known, a thermodynamic study can already enable productive preliminary material selection. This explanation can be roughly applied to non-oxide ceramic materials, too.

Both for material separation and for material synthesis, in different process stages oxide and non-oxide ceramic machines and plant components are used very successfully today in sometimes extremely aggressive conditions, which, as well as demanding chemical corrosion resistance from the materials at high temperatures, require a high mechanical and tribological loadability and often high thermal conductivity. Under such conditions, the use of ceramic components is sometimes the last resort, making economic operation of machine or plant possible in the first place.

Besides demanding resistance to the four types of load described in the introduction, the processing of active ingredients in the pharmaceuticals industry usually requires verification of suitability for direct contact with pharmaceutical products. In this field, both oxide and non-oxide materials can generally be used without any reservations. Many producers of oxide and non-oxide ceramics have obtained appropriate certification for the materials that they offer for such applications from the responsible regulatory authorities.

References

- J., Hazell, Paul (2015-07-29). Armour : materials, theory, and design. Boca Raton, FL. ISBN 9781482238303. OCLC 913513740

- Ceramics-and-glass-in-electrical-and-electronic-applications, what-are-engineered-ceramics-and-glass, about: ceramics.org, Retrieved 13 March, 2019

- The Ceramic Society of Japan (31 July 2012). Advanced Ceramic Technologies & Products. Springer Science & Business Media. Pp. 529–. ISBN 978-4-431-53913-1

- Purl, servlets: osti.gov, Retrieved 1 January, 2019

- Yang, M.; Qiao, P. (2010). "High energy absorbing materials for blast resistant design". Blast Protection of Civil Infrastructures and Vehicles Using Composites. Pp. 88–119. Doi:10.1533/9781845698034.1.88. ISBN 978-1-84569-399-2

- Thermal-insulation-application, porous-ceramics-application: induceramic.com, Retrieved 30 June, 2019

- Sur La Table; Sarah Jay (21 October 2008). Knives Cooks Love: Selection. Care. Techniques. Recipes. Andrews mcmeel Publishing. Pp. 12–. ISBN 978-0-7407-7002-9

Permissions

Index

www.ingramcontent.com/pod-product-compliance
Lightning Source LLC
Chambersburg PA
CBHW061948190326
41458CB00009B/2814